Buzz into BEEKEEPING

A Step-by-Step Guide to Becoming a Successful Beekeeper

CHARLOTTE ANDERSON
Master Beekeeper

Skyhorse Publishing

Skyhorse Publishing books may be purchased in bulk at special discounts for sales promotion, corporate gifts, fund-raising, or educational purposes. Special editions can also be created to specifications. For details, contact the Special Sales Department, Skyhorse Publishing, 307 West 36th Street, 11th Floor, New York, NY 10018 or info@skyhorsepublishing.com.

Visit our website at www.skyhorsepublishing.com.

10 9 8 7 6 5 4 3 2 1

Library of Congress Cataloging-in-Publication Data is available on file.

Cover design by Daniel Brount
Cover photos by Charlotte Anderson

Print ISBN: 978-1-5107-5739-4
Ebook ISBN: 978-1-5107-5740-0

Printed in China

For my husband, Richard, who has always encouraged me to go for my dreams.
And for my father, James Jones, who thought his little girl could hang the moon.

Contents

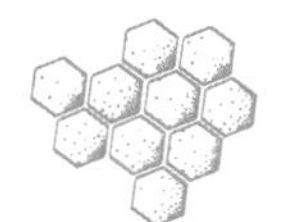

About the Author *ix*
Introduction *xi*

Chapter 1: Getting Started: Defining Your Goals **1**

Chapter 2: Challenges for New Beekeepers **7**

Chapter 3: The Life of the Honeybee Colony **13**

Chapter 4: How Bees Make Honey **27**

Chapter 5: Your First Beehives **31**

Chapter 6: Your Bee Yard or Apiary **47**

Chapter 7: The Bees Arrive **53**

Chapter 8: Finishing Out Your First Season **63**

Chapter 9: Swarming **73**

Chapter 10: Disease and Pest Control in Beehives **77**

Chapter 11: Enjoying the Bounty of the Hive **87**

Conclusion 93
Resources 95
Acknowledgments 97
Index 99

About the Author

Born and raised in the foothills of the South Carolina mountains, I always had a special love of nature. Childhood summers spent on Grandpa's farm allowed me to develop an appreciation for growing my own food and living a more self-sufficient life. My small backyard homestead features chickens, miniature donkeys, and several gardens. A few years ago, I thought that adding honeybees to the mix seemed like a great idea.

Lacking any beekeeping experience or family members who were beekeepers, my journey began. Reading, researching, attending local beekeeper association meetings, and state meetings paid off. In 2012, I became the first female Master Beekeeper in the state of South Carolina, and was named Beekeeper of the Year that same year.

Over the years, participating in numerous educational events and teaching local beekeeping classes has given me the pleasure of sharing my knowledge of bees with others. My online beekeeping class has helped hundreds of beekeepers get started each year in this fascinating hobby.

When not working in the garden, or spending time with the mini donkeys, I continue to be involved with the world of the honeybee. My website www.carolinahoneybees.com teaches others how to become better beekeepers and how to put the resources of the beehive to good use.

Introduction

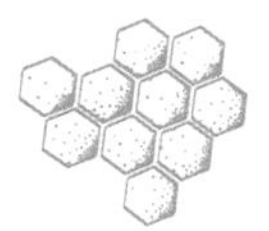

Do you envision yourself enjoying a jar of fresh, raw honey from your own beehive? If so, you are not alone. Beekeeping has become very popular in recent years. News media reports on the decline of pollinators are commonplace, spurring a heightened interest in honeybees from people in all walks of life. (Who doesn't smile when seeing a honeybee gathering pollen from flowering plants?) In wake of current concerns about bees, efforts are being made to become more "bee friendly." Considerate homeowners seek out plants and flowers that benefit bees and other pollinators, and are becoming more mindful of pesticide use, using fewer insecticides and the least toxic formulas to minimize damage to beneficial insects.

For some of us, finding simple ways to help is not enough. We want to be more involved with the world of the honeybee. We want to become beekeepers, and we want to do it for a variety of reasons. My reason for becoming a beekeeper may be different from yours. In fact, you may have several reasons to keep bees. Do you have an interest in becoming more sustainable? What about seeking a more natural life and reducing your carbon footprint? Many of us like the idea of producing our own food. When you produce your own honey, you know exactly what is in the jar. Even a small-scale beekeeper can produce enough honey for family use. In most regions, a healthy established hive of bees will

Honeybee on echinacea flower

produce at least fifty to sixty pounds of excess honey in a good year, roughly the equivalent of a five-gallon bucket. Maybe you would like to produce even more honey? If you have the time and energy, adding a few more beehives could produce honey enough to sell.

Beekeeping today is not the same as it was in Grandpa's day. Changing weather patterns affect bee health and productivity, and the introduction of new pesticides and diseases have made beekeeping extremely challenging. When foraging bees come into contact with pesticide spray, it results in bee deaths. A colony may only lose a small number of foragers to poison or the entire colony may die. Beekeepers living near large agricultural operations must be especially cognizant of pesticide poisoning. Alas, that is the way it goes. As our world changes, beekeeping methods must change with it.

Beekeeping is a combination of several disciplines. From biology to agriculture, bee life encompasses the world as a whole. It is not necessary to become a bee scientist if you want to raise bees. However, knowledge is power when managing beehives. This book is not designed to teach you every little detail about beekeeping. No one book could do that because the world of the honeybee is so diverse. My desire is to share tidbits and tips that I have learned over the years and that tend to be overlooked by new beekeepers.

The rewarding adventure of beekeeping is not without risks. You will have failures—we all do. Talk with other successful beekeepers and ask for advice, but think for yourself. As you grow in beekeeping experience continue to read and learn more about bees. Beekeeping is not a "one and done" activity. You must keep up with changes that affect your colonies. Forgive yourself for making mistakes with your bees—it happens to everyone. If you apply what you learn, your successes will outweigh your failures by a large measure.

The Bees and Me

Have you considered beekeeping for years? I did. While not lucky enough to have close friends with bee knowledge, I always found the idea of keeping

Tall beehive

wild bees in a box fascinating. And to be able to produce my own honey, that sounded incredible. This idea appealed to my sense of being more self-sufficient and knowing the origin of my food. (And the idea of selling excess honey to pay my beekeeping expenses sounded great, too!)

I recommend beginning your beekeeping journey at least six months before your bees arrive. It is a good idea to take a couple of classes because the management of honeybees involves a lot of opinions. We all have our own point of view and you will benefit from exploring different ideas. A good beekeeping class can also help with the basics. You will learn the basic parts of a hive, how to buy bees, and basic management techniques. My website, www.carolinahoneybees.com, has helped many new beekeepers. Also, check out local beekeeping clubs. They often offer classes in late winter. Beekeeping in today's environment is not easy, but you can be successful if you continue to learn and work at it.

One of the best things about beekeeping is that anyone can do it. The honeybee is the only insect in the world that produces food for human consumption. And you can have a box of these wonderful insects in your backyard. If you are thinking about a beekeeping adventure, don't wait for the perfect time. That day may never come and you will have missed the experience of a lifetime.

Having Your Own Bee Farm

A beekeeper is also called an apiarist. This term comes from the Latin word *apis* meaning *bee*. The term apiary refers to a location where beehives are kept. If you have a single beehive in your backyard, that is your apiary.

Do you dream of having a profitable bee business? Bee farming is a great goal. But first, gain some experience in hive management before attempting beekeeping as a commercial business. As you plan your own bee farm, grow slow. Starting out with a large number of beehives is a recipe for disaster. A large number of hives require a sizable monetary investment. More hives require more time (and hard work, too)! It is easy to get in over your head and end up with sickly colonies. Are you open to the idea of spending some sweaty hours in a bee suit during summer? If not, commercial beekeeping may not be right for you.

Not everyone wants a large bee business with added expenses and risks, but operating a small hobby bee business is a simpler way to let the bees pay their own way. Many beekeepers produce enough honey for their own use, then sell the rest of the honey crop. This allows the beekeeper to recover some of the costs of bees and equipment.

Chapter 1

Getting Started: Defining Your Goals

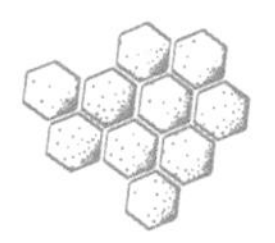

How you approach beekeeping will in part depend on what you hope to achieve. It is best to be realistic about your beekeeping goals. Do you want to produce honey for your family and friends? A couple of hives will be enough for this purpose. For some, beekeeping becomes a family project with all members involved. At most, if you have a large family and numerous friends hoping to share in the honey harvest, four hives should be sufficient.

Would you like to start a beekeeping business and sell honey to consumers? In that case, you will need more hives. Most small honey producers have at least ten to twenty beehives. The more honey you wish to produce, the more hives you need. But remember, more hives = more expense = more work. Honey production is dependent on favorable weather and varies each year. Poor bee health and management problems may result in a poor honey crop. Yes, it is possible to have ten hives that do not produce as much honey as four healthy strong beehives. Some hives will not produce extra honey each year.

Not everyone who keeps bees is interested in honey production. Bees can also be kept for pollination of fruit trees or a vegetable garden. Crop yield is increased with beehives nearby to aid in pollination. A couple of honeybee hives will provide ample pollination for a large garden or small orchard. Hives used for pollination still require beekeeper supervision. Most unmanaged colonies die within two years from pests or other health issues.

Bees eating honey in hive

As more of us seek ways to reconnect with nature, some folks want bees just for the experience. This beekeeper cares little about a honey harvest or increased crop yields. Understanding the beauty and organization of the honeybee colony is an amazing experience. This social organism (the bee colony) functions as 60,000 individuals and also as a single entity.

Determining your reason for becoming a beekeeper will help shape your start-up plan. If you only want enough honey for personal use, you don't need twenty hives! A large honey extractor or thousands of investment dollars is not necessary for small-scale beekeeping.

So, how many hives should a beginner have during the first year? You can start beekeeping with one hive but we usually advise against it. If you have two hives, you are able to share resources (eggs or brood) between them. If one of your hives fails, you still have one to work with. It is very depressing for a new beekeeper to work hard all season and have nothing to show for it. Having two hives is not a lot more work to manage than one, and doubles your chance of success.

Learning from Experienced Beekeepers

Finding a group of beekeepers to connect with is very useful. This may be a local beekeeping association or even an online beekeeping group. However, careful observation of any gathering of beekeepers will reveal a simple truth. We cannot agree on much of anything. Be prepared to hear a lot of conflicting advice. The subject of how to manage honeybees creates many differences of opinion. From which style of beehive is best to whether or not you should feed your bees, you name it and we can disagree about it. This is maddening to the new beekeeper who only wants a straight answer, but don't be too discouraged. The strong differing opinions on managing honeybees can be a blessing. It shows there are many different ways to keep bees and be successful.

Talk to beekeepers who have a track record of keeping most of their hives alive through winter. These people are doing something right. Anyone can be a beekeeping hero in the bountiful months of April and May. The true test of your skills is during times of colony stress. Summer droughts and winter cold are times when all beekeepers face difficulties. Yes, even great beekeepers lose colonies.

As you learn and listen to other beekeepers, you will want to experiment. This is a great way to learn new techniques. Your beekeeping philosophy may change over time. That is okay. Our environment is changing. Beekeeping has to evolve with the current conditions. New pests and disease problems arrive on the scene constantly. Be willing to learn and stay current on the latest beekeeping issues. A subscription to a good beekeeping magazine such as *Bee Culture* keeps you up to date with new challenges. Then, decide how you want to do things. And remember, even with the most well-meaning advice, your climate, type of honeybee, and management style all play an important role in what is truly best for your beekeeping philosophy.

Your Beekeeping Philosophy

As the beekeeper, you are responsible for your beehives. You decided to get bees and put them in your hive. All management decisions rest in your hands. How do you want to manage your bees? Will you use chemical treatments when needed? If so, which ones will you choose? Perhaps you think the bees should take care of themselves and live or die? The way in which you maintain your colonies is your "beekeeping philosophy." If you want healthy productive bee colonies, some decisions must be made.

The influx of new pests and diseases have resulted in a need to use various treatments in most bee colonies. When varroa mites first became a problem in the US, those beekeepers who did not adapt to new methods lost all their colonies. Virtually all the wild colonies of honeybees were wiped out. The old way of beekeeping does not work today.

Among beekeepers, the greatest disagreements on hive management methods revolve around controlling varroa mites. Varroa mites (*V. destructor*) are the number one killer of honeybees worldwide. Large varroa infestations weaken colonies and spread diseases. This small, reddish, external parasite is native to Asia. Bees in the native region have evolved over many years to deal with varroa, but our species of honeybee has no defense against these mites. When varroa mites were first found in the US during the mid-1980s, colony losses were massive. Something had to be done—and quickly.

A variety of chemical treatment options have been used in the years since. Some treatments are successful while others . . . not so much. There is no perfect answer to the bee mite problem at this time. Each beekeeper must make their own decision about the best way to control varroa mites.

Beekeepers generally fall into one of three groups or treatment philosophies for varroa mites. The first group of beekeepers use any mite treatment options that are approved by the United States Environmental Protection Agency (EPA). These methods have been tested and deemed "safe" to use in beehives. Some of the treatment options are synthetic chemicals. There is concern about the possibility of these chemicals contaminating beeswax or honey. We do find some chemical residues left behind after treatments, but how much is too much? There is also concern over the mites becoming

Varroa mite on bee

resistant to chemical treatments. Treatments in this group include: Apistan (fluvalinate), Apivar (amitraz), and Checkmite+ (coumaphos), among others.

The second group of beekeepers try to avoid the harsher synthetic mite treatments. They choose the "softer" chemical treatments for mite control.

Why wouldn't everyone use this treatment plan? Softer treatments are often not as effective as the conventional treatments. They may require more applications to get the mites under control. And some of them are very temperature dependent, requiring temperatures within a certain range for maximum efficiency. Treatments in this group include: Apiguard, formic acid, oxalic acid, and Hopguard, among others.

The third group of beekeepers feel you should never use any chemicals inside the hive. They believe that treatment-free beekeeping is the only acceptable management method. I understand the theory behind this approach. Could these chemicals used to treat honeybee pests hurt your bees? Will chemical residue end up in honey for human consumption? All are valid concerns., but we must remember that chemicals are already in the environment. Your bees will bring chemicals into the hive via water, nectar, and pollen. Of course, beekeepers don't want to add more chemicals to the hive. But we know that pests (like varroa mites) will kill our bees. Some plan of action must be in place. My heart agrees with the treatment-free beekeepers. However, I have not been able to make it work for me.

Most beekeepers fall somewhere in one of these three groups. You will have to decide how you want to keep your bees. In the future, we hope to breed bees that are better able to withstand mite infestations.

Treatment-Free Beekeeping

Treatment-free beekeeping has been tried by many with varying levels of success. This method of beekeeping often results in letting a lot of bee colonies die in the beginning. This can be a time-consuming and expensive process. If you feel strongly about this method of bee management, you must search out local resources. Find a successful treatment-free beekeeper in your area. Learn what methods they are using to manage honeybee health concerns. They will also be a possible source for bee purchases.

Don't try treatment-free beekeeping with bees from conventional commercial sources. The bees will die and you will spread more pests into your area. It is unfair to the bees and to your beekeeping neighbors.

Chapter 2

Challenges for New Beekeepers

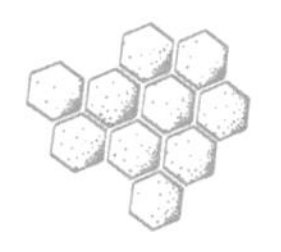

Beyond buying a hive and stocking it with bees, what else is required for beekeeping? Keeping a hive of bees is very different than putting up a birdhouse. With a birdhouse, you simply hang it up and clean it each year, easy peasy. But with beekeeping . . . that's not how it works. Your honeybees require regular hive inspections and certain maintenance procedures. Some tasks need to be performed during specific times of the year. (This will depend to a degree on your climate.) If you plan to install your bees in a hive and forget them, your chances of success are slim. Good beekeeping requires some hard work. But the reward is sweet.

Time Requirement

The amount of time required for your beekeeping tasks will depend on several factors. If you have a lot of hives, you will need to devote more time to beekeeping. However, all beehives require routine inspections. For most colonies a brief inspection (fifteen minutes every two to four weeks) during the warm season is average.

Some bee colonies will need extra attention, and you may need to feed your bees. This is especially true for new colonies or any hives during a time of drought. But you don't have to open your hives every day. Unless you are dealing with a difficult problem, an average hive inspection should take less than fifteen minutes. With prep time and cleanup added, you should estimate about thirty minutes for each hive.

All beekeeping has a local component. The weather conditions and blooming plants in one region may not be the same 100 miles away. Luckily, most beekeepers love to talk bees. Make friends with some beekeepers in your region. They will be able to provide you with the best information on local conditions. They know which plants provide the best nectar and pollen and when they bloom. Ask them how much honey should be left on the hive for winter. Be prepared, many of them will want to tell you that the way they do things is the only way. That's not usually true, but don't miss the opportunity to learn from their experience.

Beekeeping Is Hard Work

New beekeepers are exposed to a lot of information. Feeling overwhelmed is common. You may be filled with doubts regarding your ability to keep bees. Can you do this beekeeping thing? Yes, most folks can be successful with a beehive. Don't be frightened by a long list of bee diseases and problems that you will probably never see. However, you do need to know how to recognize problems in the hive. Some hive issues resolve themselves but others require beekeeper intervention. Learn how to recognize problems that may arrive in your hive. If you are concerned about disease, your state bee inspector is a valuable resource. Reach out to your state department of agriculture for contact information.

Beekeeping has some physical demands but most difficulties can be overcome. If you are concerned about being able to lift bee boxes, find a workaround. As a woman with limited upper body strength, I can still move a full hive. I simply lift one box at a time. If a honey box is too heavy to lift, transfer half the frames into an empty box and divide the weight, then put the frames back in place after you finish your inspection. This is another good reason to find a local beekeeping buddy. You will have someone to help with the heavy lifting. (And having a friend to take to bee meetings is a lot of fun!) But be forewarned, if you are afraid of sweating, beekeeping may not be for you. Many of the hive management tasks must be done during the hot summer months. We southern beekeepers are all too familiar with a sweat-drenched beekeeping suit.

Bee Stings: It Happens

The possibility of being stung by your bees is a reality. And being concerned about bee stings is only natural. Will you get stung while working your beehives? Yes. Working with honeybees does involve some stinging situations. You should not get stung every time you inspect your hives. However, it does happen on occasion.

Understanding why bees sting doesn't make the pain any less. But it can help you understand why stings happen and how to avoid them. Honeybees do not sting out of anger—though it may seem that way sometimes. Worker bees have the task of defending the hive from predators. Your bees cannot know that you mean them no harm. You are viewed as an intruder. When a threat is detected, the bees serving as hive guards release alarm pheromones, chemical messengers that are similar to hormones except they are on the outside of the bee's body. This alarm pheromone calls other bees to come to the defense of the colony. Dozens of bees fly out to aid in the attack. Finding a giant dressed in a white bee suit, they do not pause to question your intentions.

Getting the Stinger Out

No matter how careful you are, inevitably you will be stung. When a honeybee stings you, her barbed stinger gets stuck in your skin. The stinger and poison sac rip from the body of the honeybee. Alas, she will die. She has given her life for the colony. And you are left with pain and discomfort.

If a stinger is embedded in your skin, you must remove it quickly. Do not grab the stinger with your fingers. This will only force more poison into your skin. Instead, scrape the stinger out of your skin. A quick scrape with your fingernail or hive tool will get the job done. Luckily, most of us will only experience a normal reaction. Some pain, redness, and a little swelling and itching is common for those of us not allergic to honeybees. It should clear up in a day or so. A cool ice pack applied to the sting site and maybe some antihistamines can offer relief.

More Experience = Fewer Stings

Experienced beekeepers tend to have fewer stings. This is due in part to learning when to inspect your bees. During midday, the older worker bees will be busy foraging. This results in fewer guards being in the hive, and fewer opportunities to get stung. Weather also plays a role in hive temperament. Inspect your hives on sunny, mild days if possible. A weather front moving in tends to set the bees on edge. Can you blame them?

When performing hive inspections, try to avoid quick, jerky movements and don't bang the hive components together. Move your arms slowly, and don't breathe on the bees! Stand behind your hive during hive inspections. This keeps you from being in the way of bees that are returning to the hive. Bees are very sensitive to odors so make sure you're not emitting a strong smell—good or bad. These techniques will lessen your chances of getting stung while working the colony.

Learning how to use a bee smoker reduces stings and saves bee lives. This essential beekeeping tool is not harmful to the bees. When used properly, cool white smoke disrupts the alarm pheromones of guard bees. Gently puff a little smoke at the hive entrance. Wait a couple of minutes before opening the hive, then proceed with your inspection. Periodically during your inspections, a little puff can help keep the bees calm.

Everyone wants to feel comfortable while working with their bees, but I always recommend wearing beekeeper protective gear. You may not wish to wear a full beekeeper suit but a proper hat and veil should be a minimum, unless you are willing to put your eyesight at risk. A sting in the eye could result in vision loss.

Are You Allergic to Bees?

Many people believe they are allergic to bees, but most of us do not have a bee allergy. If you are allergic to wasp stings such as yellow jackets you may or may not be allergic to honeybee stings. The venom of honeybees contains different proteins than wasp venom, but you may still be allergic to both honeybees and wasps. A local reaction to a honeybee sting is experienced by most people, and includes some pain, swelling, and itching that resolves in a day or two. This does not mean you are allergic.

However, for about 7 percent of the population, a bee sting becomes a more serious matter. These people have a true allergy to bee stings. A systemic reaction spreads all over the body, affecting the skin and respiration system. Emergency treatment is required for difficulty breathing, hives, facial swelling etc. Call 911 if you suspect a dangerous allergy to bee stings.

Can people with bee allergies become beekeepers? Sometimes, but it depends on the severity of the bee allergy. I have a friend who is allergic and she is a beekeeper. She takes extra precautions. She wears additional clothing under her bee suit and heavy gloves. She understands and assumes the risk and of course keeps epinephrine close by.

Beekeepers can also develop a bee allergy over time. Some people keep bees for years and then become allergic. Any sting reaction beyond mild pain, swelling, and redness should be evaluated by medical personnel. If you suspect a possible bee allergy, get checked out by a physician before keeping bees. And if you notice a change in your reaction to stings, have a serious conversation with your health professional.

Chapter 3

The Life of the Honeybee Colony

Honeybees Are Social Insects

Honeybees are true insects. They have three main body sections (head, thorax, abdomen), six legs (three pairs on each side), and four wings (two pairs on each side). A honeybee colony is a group of eusocial insects, or social insects that live and work together for the good of the whole. It fulfills the three requirements for any social group:

- Cooperative brood care
- Reproductive division of labor
- Overlapping generations

Inside the beehive, cooperative brood care is observed. Worker bees are females. They take care of larvae that are not their own offspring. The tasks of reproduction and housekeeping are shared by all the workers in the colony. Not every female honeybee will reproduce but all will help feed the baby bees. With the exception of the queen, individual bees do not live from one year to the next, but with good health and proper conditions, overlapping generations will always be active in the hive. Being an expert on honeybee biology is not a requirement for becoming a good beekeeper, but good hive inspections are impossible without being able to identify the different bees and know their role in the hive.

Honeybee Taxonomy

The list below illustrates the bee family tree for those of you who are scientifically inclined.

Kingdom—Animalia: Honeybees are animals.
Phylum—Arthropoda: They have exoskeletons and jointed legs.
Class—Insecta: They are insects with three-part bodies (head, thorax, abdomen).
Order—Hymenoptera: This means "membraned wings." Bees, ants, and wasps belong here.
Family—Apidae: This means bee.
Genus—Apis: This means bee, too!
Species—Mellifera: Honey producer.

Worldwide there are many different species of honeybees, but in the United States we only have one species of honeybee: *Apis mellifera*. Species do not interbreed. However, races of honeybee within a species can breed. Common races of honeybee include Italian (*Apis mellifera ligustica*), Carniolan (*Apis mellifera carnica*), Caucasian (*Apis mellifera caucasica*), Africanized (*Apis mellifera scutellata*), and others. It is unlikely you will be able to find a purebred bee. Most beekeepers have mutt bees that are a combination of genetics from several races. In the past, Buckfast bees (a hybrid) were popular and they are still available. Today, more emphasis is being placed on hybrid bees with hygienic traits that may be more tolerant of varroa mite infestations. Hygienic bees are a mix of several different races.

Africanized Bees: Killer Bees

A quick reference to the bees referred to as "killer bees" or Africanized honeybees (AHB). They are a cross between a race of bees found in Africa and regular European honeybees. AHB escaped from a breeding program in South America, then expanded north. They now occupy many southern states in the US. Africanized bees are extremely defensive and are not suitable for regular beekeepers.

The Bees in Your Hive

One of the greatest joys of beekeeping is looking inside the hive. Remove a frame and you'll see thousands of busy bees moving around on the comb. You may have workers of various shades in the colony because they have different fathers. Beyond color variations, bees in your hive have other differences. A honeybee colony is home to three types of bees. The number of each type of bee will vary with season and foraging conditions. However, each kind of bee has an important part to play in the colony. One of the first tasks for any new beekeeper is learning how to recognize the different bees.

Three Types of Bees in a Summer Beehive

- Drone bees (male) number into the hundreds during mating season
- Worker bees (female) number into the thousands
- Queen bee (female) usually makes up one only per colony

Each honeybee colony consists of thousands of individual bees. An average hive boasts a population of 30,000 to 60,000 members. A productive bee colony needs a large population of worker bees. During the warm months, bees work to collect nectar and make honey. Without enough stored honey, the colony will starve before spring. Winter survival depends on the colony having enough food stored. They also need a population large enough to maintain warmth during the long cold months.

Colony population fluctuates during the year. Late winter/early spring colonies are at the smallest in number of bees inside. But healthy overwintered colonies grow quickly once nectar and pollen become available. Warm weather, increasing length of daylight, and available nectar prompt the queen to increase egg production. In perfect conditions, colony population grows so quickly that swarming may result. Colony strength usually reaches its peak in June through July. From midsummer to fall, colony population stabilizes and then slowly drops as winter arrives.

The Honeybee Life Cycle

The actual life span of a honeybee varies with the type of bee and other hive conditions. However, as an insect, honeybees go through four life cycle stages.

- Egg
- Larva (milk brood)
- Pupa (capped brood)
- Adult

Every honeybee goes through these four stages, but the development time to adulthood is different for each kind of bee. The beginning of the honeybee life cycle is the egg. A mated queen bee lays one egg in each cleaned beeswax cell. Below is the time required for that egg to become an adult. These numbers can vary slightly in hot weather.

- **Drone bee:** 24 days
- **Worker bee:** 21 days
- **Queen bee:** 16 days

Understanding the role of each type of bee is important to good hive management. This is why beekeepers must learn how to identify each kind of bee. A summer colony with thousands of drones and few workers has a problem. Seeing drone production in late winter hives is a sign that the bees are getting ready for spring. Drones are necessary for reproduction and fertilization of virgin queens. During any inspection you should see mostly worker bees and worker brood.

The Drone Bee

Drone bees are the only males in a hive. They develop from unfertilized eggs! Yes, a queen does not have to be mated to produce drones. This is known as parthenogenesis, the ability to reproduce without needing both egg and sperm. The egg develops without being fertilized, and takes twenty-four days to develop from egg to adult. This is the longest development time of any of the bees. A reasonable number of drones (several hundred) is normal in a healthy bee colony during the warm months. Beekeepers sometimes consider drones a liability or drag on colony resources because they do not forage for food or protect the hive. However, drones are important to colony balance and serve a very important role as we soon shall see.

Drones are noticeably bigger than worker bees. In fact, they are often mistaken for a queen bee by beginner beekeepers. (Well, until the beekeeper notices that there is another, and another, and another.) Drone bees do not have stingers; their blocky bodies have rounded fuzzy tails instead of coming to a point at the end. They do have a very large pair of compound eyes that cover the top of their head. These large eyes tell us something about the purpose of drones. They need to be able to see virgin queens flying in the

Drone bee with large eyes

air. The sole purpose for drone bees is reproduction. Upon maturity, drones leave the hive each day looking for virgin queens. These mating flights take place on warm afternoons. The drones that are successful in mating with a queen will die shortly after. The unsuccessful male bees (that's most of them) return to the hive to eat and rest. They try again on another day. This process continues all summer if foraging conditions are good.

Drone bees live for several months except for those that successfully mate and die. But woe to those unlucky fellows still alive as cool weather approaches. When fall arrives, the workers force the drones out of the hive to die. Why? The colony does not need drones during winter when no new queens need to be mated. So, why feed them? Very practical are honeybees, if a little harsh.

The Worker Bee

Worker bees are non-reproductive females and represent the largest population in the colony. Thousands of worker bees do the tasks that keep the colony fed and safe. During summer, a colony of bees can have 30,000 to 60,000 bees in the hive. During winter, the population of the colony will be much less. Exactly how big the winter population is depends in part on the genetics of the colony. Italian honeybees tend to have larger winter populations. Russian bees and Carniolans keep smaller populations over winter.

The worker honeybee has several specialized body parts:

- Brood food glands in the mouth
- Wax glands on the underside of the abdomen
- Pollen baskets
- Honey stomach or crop
- Barbed stinger

Brood food glands are located in the mouth of the worker bee. These glands secrete food for growing larvae. Royal jelly and other food compounds promote rapid larval growth. It is the job of young nurse bees to feed the baby

Nurse bees at work raising young

bees. When working as a nurse bee, the brood food glands will be large and fully functional.

In addition to feeding baby bees, worker bees also produce beeswax. Beeswax has several uses inside the hive. In fact, it forms the foundation of the colony. Young adults are the best wax producers. Special wax glands on the underside of the worker bee's abdomen secrete soft wax scales. These are shaped to form sheets of honeycomb with thousands of beeswax cells. Beeswax production requires a lot of colony resources. Workers must consume a lot of food to activate their wax glands. A well-fed colony with a lot of young bees is an efficient comb-building colony.

On the hind legs of the worker bee we find a patch of stiff bristly hairs; these are called pollen baskets. As the bee gathers pollen from flowers, she

Stored pollen in comb

wets it with saliva. It is then pushed onto the pollen baskets and brought to the hive. Back at the hive, the forager finds an empty cell and deposits her pollen load. House bees mix enzyme-rich saliva with the pollen to create "bee bread." Fresh raw pollen would spoil. Bee bread can be stored for months and used as needed. Pollen is very important to a honeybee colony. It is not used to make honey. However, no baby bees can be raised without this source of protein.

Of course, we know that honeybees make honey. A special organ inside the worker bee abdomen allows her to transport nectar. The "honey stomach" or crop is located between the esophagus and true stomach. This pouch holds nectar as it is collected. When the honey stomach is full, the bee returns to the hive and transfers the nectar to a house bee. House bees are young adult workers who have not yet begun to forage. They will complete the process of transforming nectar into honey.

Another special structure of the worker bee is her stinger. The stinger is a valuable tool for a worker bee who is responsible for defending the colony. The barbed stinger and attached poison gland make it possible for a bee to deter a larger predator. However, the bee pays a big price for this advantage. Honeybees that sting mammals usually die shortly after.

Role of the Worker Bee

Worker bees develop from fertilized eggs. Emerging from their cells twenty-one days after the egg is laid, they get to work immediately. The role of the worker bee is filled with variation. Different tasks are performed depending on the age of the bee and needs of the colony. Bee jobs include: cleaning and polishing wax cells so the queen can lay a new egg, feeding bee larvae, making beeswax, producing honey, tending to the queen, helping cool or heat the hive, guard duty, foraging, and others. The time spent at each task can vary according to the needs of the colony. However, most workers follow a general progression of tasks. The first three weeks of the worker bee's life are spent working inside the hive. During the last three weeks, our worker bee becomes a forager. She goes into the field and brings home nectar, pollen, or water. Some bees will collect plant resins that will become "propolis," a bee glue used inside the hive.

Worker bees are aptly named because during the summer they work themselves to death. Gathering nectar from flowers and bringing in pollen is hard work. After weeks of foraging, bee wings and body parts begin to wear out. Bee anatomy does not allow for replacement of worn body parts. Once the wings are too tattered to fly, the bee is done. She may die inside the hive and be removed by other workers, or not make it back to the hive at all.

Worker bees have the sole responsibility of preparing the colony for winter. The life span of a busy worker bee is only about six weeks during summer. Therefore, a constant supply of new adults is required. If a bee colony does not have a productive queen bee and ample worker bees to raise young, the population will drop. During any hive inspection, always verify the presence of a laying queen with a good pattern of worker brood.

As summer wanes and winter approaches, a change takes place in worker brood. Worker bees produced in the fall can live up to six months because they are physiologically different and have larger bodies that help them survive over winter. Beekeepers must manage their colonies to ensure good health in late summer so that healthy well-fed nurse bees can raise strong worker bees to last through the winter. If sick, pest-infected colonies are not able to rear healthy, fat winter bees, when fall arrives the colony may not be ready for cold weather. The result may be the death of the entire colony before spring.

The Queen Bee

The queen honeybee is a female. She develops from a fertilized egg to adult in sixteen days, the shortest development time of any bee in the hive. She

The queen bee

is the only reproductive female in the colony. This means she is the only bee who is able to mate and lay fertilized eggs. Only fertilized eggs produce worker bees. In normal circumstances, there will only be one queen bee in a hive. The exception to this is when a mother/daughter queen coexist for a short time. But eventually, the new queen will take over. The queen is the largest bee in the colony. Her midsection or thorax is slightly larger than that of a worker bee. Due to her long-tapered abdomen, she is able to lay eggs in the bottom of honeycomb cells. The queen does have a stinger but it is smooth; not barbed like the stinger of worker bees. Why is her stinger different? Her stinger is used to combat rival queens. When a colony needs a new queen, several new queens are produced at one time. The strongest queen kills the other challengers. Even though the queen bee can sting, it is rare for the beekeeper to be stung by her.

A queen is the only female honeybee that is able to mate. Honeybee mating does not take place inside the hive. A virgin queen must leave the hive and find drones to mate. A few days after emerging from her cell, the virgin queen takes flight on a warm, fair afternoon. She flies to a special place known as a "drone congregation area" seeking drone bees. How do the bees know where to go? We don't know the answer to that question . . . but the bees do.

Mating takes place in flight with the drone mounting the queen in mid-air. Once mating is accomplished the drone falls to the ground and shortly dies. The queen makes several mating flights and mates with twelve to twenty drones over the next couple of days. Semen is stored in a special organ in her abdomen called a "spermatheca." After this time of mating is completed, the queen never mates again. Once her reserve of stored semen (sperm) is depleted, she will only be able to produce drones. We call this a failing queen and the colony will attempt to replace her. Other than mating flights, the queen never leaves the hive again unless she accompanies a bee swarm.

The queen bee fulfills another special purpose. In addition to egg laying, the queen bee produces pheromones. These external hormones (chemical messengers) are very important to the honeybee colony. Pheromones serve as a communication system that governs colony activity. A queen bee with declining egg production or declining pheromones is a bad sign. This signals the colony to create a new queen bee. It is rare indeed to have a productive queen more than two years old. Beekeepers often replace aging queens.

How a Queen Is Made

Queens may be the longest-living bee in the hive but no bee lives forever. When a queen bee dies, the colony must make a new queen quickly. Luckily, they have a plan for this that works very well most of the time. Several very

Queen cells on frame

young larvae (from fertilized eggs) are selected. These larvae receive a special diet of royal jelly and rich food substances in ample quantities that cause the larvae to develop into sexually reproductive queen bees. Any young female larva has the potential to become the colony's queen bee. Queen larvae grow much larger than regular worker larvae and need more room to develop. A regular sized honeycomb cell just will not work, so workers build large peanut shaped cells around the developing queen larvae. We call these queen cells. On or around the sixteenth day of development, the queens will hatch or emerge from their cells. Several queen cells are produced during the re-queening process. The first queen to emerge seeks out and kills her sisters. It is a fight to the death, with only one queen remaining to rule the hive. Within a few days, she leaves the hive to mate and returns to become the new queen.

During the warm season, an aging queen bee may leave the hive with a swarm. Swarming is the honeybee way of reproducing on the colony level. When ready to swarm, a queen and about 50 percent of the population will leave to form a new colony. This leaves the original hive with queen cells to make a new queen bee. If the old queen is failing, the new swarm hive will replace her soon after establishing their new home.

Occasionally a swarm will have several virgin queens. We beekeepers may not notice because we find one queen in the swarm and think, *there she is*, then we stop looking. Once the swarm arrives at their new location, one of the virgin queens will mate and become the mother of the hive.

Chapter 4

How Bees Make Honey

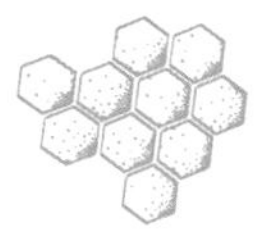

As we have learned, worker bees do all the foraging for the colony. Traveling from flower to flower, nectar is collected using her proboscis (a long straw-like mouth part). A nectar forager visits many flowers before returning to the hive. Nectar is stored in her crop, or "honey stomach." Located directly above the natural stomach, the honey stomach is not part of the digestive system, but if the forager is hungry, she can open a special value and allow some nectar into her real stomach. During nectar collections, the worker bee adds a little saliva to the nectar, making it easier to pass to her honey stomach. When her honey stomach is full, she will return to the hive with her bounty. She may travel with a load of nectar or pollen that is almost equivalent to her own weight.

Upon reaching the hive, the forager bee opens a small slit in the top of her proboscis. This allows her to share the nectar with others. House bees and other foragers can enjoy a sample of her hard work. When her honey stomach is empty, she is ready to return to the field and gather more.

House bees complete the process of making honey. They manipulate the watery nectar with their mouths, adding enzymes and reducing water content. The enzyme "invertase" is added to the diluted nectar. This enzyme is produced by the hypopharyngeal glands in the mouth/head of worker bees to change the chemical composition of the nectar, along with glucose oxidase (another enzyme), and bee saliva. Sugar molecules are then created and begin to change form, producing gluconic acid and hydrogen peroxide. These give honey its acidic and antibacterial properties. The last step in making honey is dehydration. Bees release a drop of nectar on their mandibles

Bees capping honey

(jaw). Exposed to the warm, dry air inside the hive, the moisture content begins to drop. Nectar in the drying (or moisture reduction) process is placed in droplets along the surface of comb. House bees fan their wings to increase air flow through the hive. This aids in lowering the water content of the nectar. When the moisture content of the nectar has dropped from about 70 percent to 20 percent, we consider the transformation to be complete. The bees have made honey from plant nectar.

Honey is more stable than watery plant nectar and less likely to spoil. Finished honey is stored in beeswax cells and sealed with a cap of beeswax. When beekeepers harvest excess honey, they take honey that is capped with beeswax because it signals that the nectar to honey conversion is complete. The honey is ripe and ready to use.

When Bees Make Honey

While some species of bees do make honey, none produce large stores of food for winter like our honeybees. Bumblebees make a small amount of honey and store it in "honey pots" inside their nests. But they are not true honeybees. Only bees in the scientific genus *Apis* are true honey producers. They have evolved to make honey in time of plentiful food sources. Then, the stored honey is used for survival when collecting food is not possible.

Many people are surprised to learn that most honey production does not occur year-round, at least for bees living in regions with several cold months. Bees must have nectar producing flowers in order to make true honey, but not many plants bloom during the cold season. Even if they did, our cold-blooded honeybees could not fly to collect it, so honey production is limited to the warmer months of the year in most locations. As a beekeeper, whether or not you harvest honey once a year or more depends on your local climate. Some regions have more periods of heavy nectar production than others. We call this the "honey flow" because it is the time when bees have ample nectar to produce excess honey. In my region, we have a heavy spring honey flow and very little the rest of the year. This information is important to good beekeeping because it can help you understand when to harvest excess honey without affecting the winter food stores of the colony.

Chapter 5

Your First Beehives

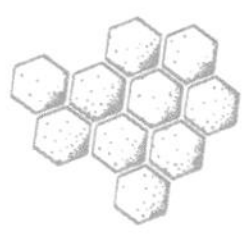

Once you have a clear idea of what you hope to achieve through beekeeping, you are ready to begin. Learning how to inspect a hive, how often to inspect, and what to look for are skills that require time to master. How many hives do you need? Beginning your bee adventure with two beehives works well for most people. Starting with only two hives (not ten) allows you to learn and grow at a calm pace. If you acquire more hives than you can manage, that is a recipe for disaster. The end result is usually many dead bees and a frustrated (former) beekeeper. A high percentage of new beekeepers leave the hobby after the first year, no doubt due in part to not understanding what they got into at the start.

I started slow with two beehives my first year. Over time, I developed what we call "bee fever" and worked my way up to about twenty-five hives. The more the better, right? Wrong. Managing twenty-five hives by myself in early spring was hard work. Working those same hives alone in the July heat of the south was a nightmare. I was not the best beekeeper that I could be because I had too many hives to manage. Know your limits, physically and time-wise. Get this one fact straight: having a lot of hives does not make you a better beekeeper. It is better to have four healthy, well cared for hives than thirty poorly managed colonies. Scorching heat or vacation plans mean nothing to the honeybees. Thankfully, most management sessions can be scheduled around the heat of the day or your week at the beach!

Buying Beekeeping Supplies: Used or New?

Beekeeping is not an inexpensive hobby. Your first year will require the largest investment of money. You will need to buy equipment for the bees, beekeeping tools, and the bees themselves. Your equipment, tools, and beekeeper protective wear should last for years. I have equipment that is over ten years old and still in fair shape. After this initial cost, expect to buy maintenance supplies such as sugar (feed) and medications. Additional beekeeping equipment is needed as your colonies grow and need more room. It does not take long before you start wondering *where am I going to put all this stuff*?

Because of the first-year expense, the option of buying used equipment looks tempting. Is used beekeeping equipment a good deal? Not always. A good deal on a pre-owned beekeeper hat and veil or suit may be okay. Used protective wear or tools can sometimes be found and a well-cleaned suit or smoker poses low risks. However, new beekeepers should avoid purchasing used beehives, frames, or beeswax comb. Diseases can be present on used equipment. American foulbrood is one disease that can live in old equipment for over fifty years. The savings is not worth the risk. Used beekeeping equipment should only be considered when you know the beekeeper. A successful local beekeeper with experience may desire to downsize his colony numbers. If they have a reputation for being a good beekeeper, it may be worth the risk. Never purchase used equipment that you know nothing about.

Which Hive Style Is Best for You?

What kind of beehive do you envision for your bees? There are several different styles of beehives used around the world. Each hive type has different management techniques. The two most common beehives in use in the US are the Langstroth hive and the top-bar hive. Successful beekeeping can be accomplished with either hive.

The Langstroth Hive

The Langstroth hive has been the commercial standard for many years. This is the common hive of stacked boxes that you see along country roads. Developed by Rev. L. Langstroth in the mid-1800s, it was the first modern hive featuring frames that could be removed. This made hive inspections possible without destroying the comb. The Langstroth hive is built on the concept of bee space. In his book, *The Hive and the Honey-Bee*, Rev. L. Langstroth explains that honeybees keep special spacing inside their hives. Even wild colonies prefer this method of spacing in the hive. Bees leave a space of approximately ⅜" between individual combs and between the combs and the sides of the hive. This bee space creates a pathway to maneuver inside the colony.

A Langstroth hive consists of a top and bottom with stacked rectangular boxes of varying depths. These boxes (also called "supers") are stacked on top of one another as the colony expands. Langstroth hives are used by commercial beekeepers. They are easy to move from place to place and stack on the back of a truck. They are also the most common style used by hobbyists and my recommendation for beginners. They are available in ten-frame and eight-frame sizes. The following are average weights associated with Langstroth equipment: a deep frame of honey weighs ten pounds, a medium frame of honey weighs seven pounds, and a shallow frame of honey weighs five pounds.

A Top-Bar Hive

Another type of honeybee hive popular in some parts of the US is the top-bar hive. The top-bar hive, also called the Kenyan hive, is one of the oldest types of beehives. It is a horizontal one-story hive with wooden top bars and no frames. Bees build their own comb hanging down from wooden top bars. No beeswax foundation sheets are needed.

Bees kept in top-bar hives will produce honey, but the top-bar beekeepers I know do not produce a lot of honey. They are keeping bees for beneficial pollination. The top-bar beekeeper who harvests honey must crush the comb and strain the honey. A honey extractor

Crush-and-strain honey harvesting

is not compatible with top-bar comb, which means that the bees have to rebuild new honeycomb each season. It is good for the bee colony to have fresh comb in the hive, but the energy required to build new honeycomb will reduce the honey harvest.

What is the perfect hive and which one should you choose? The perfect hive is the one that will work in your climate for you and your bees. What are the beekeepers in your region using? There is a lot more information available for managing Langstroth hives, but it is possible to be a good beekeeper with either Langstroth or top-bar hives.

If you want to use a different hive type, it is essential that you find successful beekeepers who are also using it. While the bees themselves may be the same, different hive styles can require a change in management practices. I often advise new beekeepers to begin with a Langstroth hive and then experiment with other types in the years to come.

Getting Bees for Your Hive

You cannot become a beekeeper until you have bees. There are several ways to get bees for your new hive. Some new beekeepers wait for the season to warm in hopes of catching a swarm. This is an inexpensive way to begin but not something you can count on. Most new beekeepers choose to buy bees. These are the most common ways to acquire bees.

- Buy a package of bees
- Buy a nuc of bees
- Buy an established full-size hive
- Catch a bee swarm

Package Bees

The most common way to acquire bees in the US is to buy a bee package. A three-pound package of honeybees contains about 10,000 bees and a queen. They arrive in a small wooden box with screen sides. All three types of honeybees will be inside the package including workers, drones, and a queen. This is a sufficient number of bees to start a new colony. The queen bee is

Package bees in a box

inside a small queen cage (with a few attendants) that feed and groom her. She is not the queen that was living with your bees. Package bees are shaken from several hives and mixed together. Your bees do not know this queen yet. They will get to know her as they slowly eat through the candy plug and release the queen from her cage. Your package of honeybees will experience less stress if you pick them up from a supplier. Let's face it, if someone put you in a box and shipped you through the mail service, it would stress you out too! Package bee shipping is available from many bee supply stores. If you are lucky enough to have a bee supply within driving distance, pick up your bees. A day trip to the bee pickup point can be a great family outing, but be sure your passengers are okay with riding home with a package of bees in the car.

It may sound strange, but be sure to keep those bees cool on the way home. Unless it is very cold, you don't have to worry about keeping the bees warm. However, they will overheat easily because they cannot spread out on comb to cool themselves. Place the package of bees in your car with a little air conditioning on or have the windows down. In the bed of a truck, keep them close to the cab to be out of the wind.

Don't be alarmed if you have some dead bees in the bottom of the box. This is normal. Bees die every day. If you have more than an inch of dead bees in the bottom, that may not be a good package. I would ask the supplier if I could choose another package, and install your bees in their new hive as soon as possible. Holding the bees in the package for additional days can be done if necessary but it adds stress.

Package bee orders are taken beginning in November through December. Yes! Order your bees during winter for spring delivery. In South Carolina, beekeepers want to get their bees early to catch the early bloom season. Those first package dates sell out so be sure to order early. If you missed the early order dates, you may get lucky and still be able to get bees, but once summer arrives, you will most likely have to wait until next year.

Nuc or Nucleus Colony

What is a bee nuc? A nuc (pronounced like "nuke") colony of bees normally consists of five frames of bees, comb, brood (baby bees), and a queen. This is the heart of a colony. A nuc is a started colony that has everything needed to grow into a productive hive. They only require space, food, and time. Nuc bees are often placed into a ten-frame box (with five more new frames, of course) to fill in the empty space. This colony is ready to grow.

There are advantages to purchasing a nuc instead of a bee package. The nuc has a head start on growth. Full frames of drawn comb are in place. The starter hive has brood that will be emerging soon to build the population and some stored food. Because of the presence of brood, nuc bees are less likely to abscond or leave the hive. The queen in the box is already known to the colony. We do not have to worry that the bees will not accept her.

Buying nucs does have a few disadvantages. They are more expensive than package bees. Nucs are commonly available later in the bloom season, though availability is limited in most regions. The biggest issue to consider when buying a nuc or any colony with comb is the risk of disease. When beekeepers refer to a "nuc," they usually mean the actual five frames of bees, brood, and queen. A nuc is also considered a small beehive box that holds only five frames. Tricky, aren't we?

A Fully Established Hive

A few lucky beekeepers may be able to find a complete hive for sale. This is a great option for those who have missed the early package bee delivery dates. Purchasing a full hive can give you a honey harvest sooner. However, this option has the same risks as nuc colonies. They cost more, have a potential for disease, and are not always available. Also, managing a full-size defensive hive can be intimidating for the new beekeeper. For this reason, I recommend saving yourself a few frayed nerves and start with a package of bees or a five-frame nuc. In general, smaller colonies are less defensive. This gives you time to become comfortable with hive inspections before the colony grows large and intimidating. As you grow as a beekeeper, you will learn more about bee math, and with experience, estimating hive strength becomes easier. A fully covered deep frame holds about 1,750 bees per side, and one pound of bees equals about 3,500 bees, so your three-pound package contains roughly 10,000 bees.

Catching a Swarm

Swarming is a natural honeybee activity. When a swarm occurs, about half of the hive population and the old queen leave the hive to make a new home. The original hive is left with queen cells, one of which will become their new queen. The mass of swarming bees makes a loud buzzing noise as they whorl in the air. It is quite thrilling to watch a swarm of bees in flight. Often, they will set down in a nearby tree for a few hours

Bees swarm in a tree

before taking off for the new nest site. This may be your best chance of catching them.

Beekeepers love to see swarms. Well, as long as it came from someone else's hive. Swarming reduces the honey production of a hive. Why? Because half of the workforce just flew away! Catching a swarm is a lot of fun and one way to get started with free bees. Check out some of the fun ways to trap bee swarms on my site (www.carolinahoneybees.com). And, of course, the Internet is full of many crazy and inventive ideas.

If you have a new package of bees, they will *probably* not swarm the first year. I say "probably" because you almost always have to say that with bees. They seem to enjoy doing what we don't think they will do! If your first-year hive swarms, please try to catch them.

Beekeeping Supplies and Equipment

Your honeybees are ordered and you can't wait for them to arrive. Don't sit back and get too comfortable; you have a lot to do! You don't want to be running around like a wild person on bee pickup day. You can save your sanity by planning ahead and being ready.

First, you must get a hive ready for them. Order your hive components (discussed on page 39) at least eight weeks before your bees are due to arrive. Bee supply companies get very busy during bee season. It is not uncommon to have orders delayed by several weeks, but the hive needs to be assembled and painted well before bees arrive, so I recommend ordering enough equipment to get your bees through their first year. I will concentrate on the Langstroth hive (available in ten- and eight-frame sizes) because I think that is best for beginners.

Some beekeepers like using a Langstroth hive that holds only eight frames. This type of hive is easier to lift and move. Those beehive boxes get heavy! There always seems to be a trade-off in beekeeping. The eight-frame equipment is easier on the beekeeper's back but it also leaves less room for the bees to grow. And, a colony may swarm more often in eight-frame equipment. More frequent inspections and making sure the bees have plenty of space can help reduce swarming.

Langstroth Hive Components

- Bottom board (screened or solid wood)
- Deep hive body with frames and foundation inside (eight- or ten-frame)
- Shallow super (or a second deep or medium in cold climates to hold winter honey for the bees) with frames and foundation
- Inner cover (provides insulation and keeps frames from sticking to the top)
- Telescoping top (protects the colony from weather)

The standard Langstroth hive box measures $19\frac{7}{8}$" in length and $16\frac{1}{4}$" wide. Boxes or supers are available in three heights:

- Deep or hive body super: $9\frac{5}{8}$"
- Medium or Illinois super: $6\frac{5}{8}$"
- Shallow super: $5\frac{7}{8}$"

How many boxes do you need for your bees? Some beekeepers use two deep/hive bodies for their bees' brood area and winter food. Anything above that will be the honey harvest for the beekeeper. The needs of your bees will depend on your climate. My standard hive configuration is: one deep/hive body and one shallow. In these two boxes the bees store food and raise young bees. Honey supers (shallow boxes) will be added on top for my honey harvest.

Inside each box are frames (usually wooden) that hold sheets of foundation. The purpose of foundation is to encourage the bees to build straight comb. This eases frame removal during hive inspections. Beekeepers can choose plastic foundation or beeswax foundation. I prefer beeswax foundation that has small wires embedded. The support wires make the frame of comb stronger, and support wires in the foundation allow beekeepers to use an extractor for honey harvest. Then, the empty frames of comb can be reused by the bees. In the past, beekeepers would cross wire these frames of foundation. That method has fallen out of use in recent years but the supplies are still available from bee supply stores if you wish to give it a try.

Bees building comb

Choosing the best hive configuration for your hives may require some testing. The exact size and number of boxes you use will depend on your climate and beekeeping philosophy. What do the other beekeepers in your region use? Common setups for a Langstroth hive are one deep and one shallow; one deep and one medium; two deeps; or three mediums.

New hives or small swarms with low populations may be unable to defend a fully opened entrance. Use an entrance reducer to make the hive opening small. This wooden strip fits in the open space between the bottom board and the hive body. It has different sized openings that can be used by rotating the reducer. Open the entrance to a larger size as the colony grows.

Another important hive accessory is a bee feeder. There are many ways to feed bees, including the popular quart jar method. Small holes in a quart jar lid allow bees to drink sugar water from the inverted jar. It is often placed inside the hive with an empty box around it. You can also feed bees outside in an open feeder. This is a less efficient method because you will also be feeding every wasp and yellow jacket in the neighborhood, and your weaker colonies will not benefit as much because they have fewer workers to collect syrup. In addition to jar feeders, you will find bee feeders in many styles. All of them have advantages and disadvantages. Choose one that appeals to you and test it with your bees. Feeding greatly increases their chances of thriving.

Failure to feed new packages properly is the most common mistake made by beekeepers in my beginner classes. Bee syrup is easy to make and can be stored in the refrigerator for a week or two. To make sugar water for a new bee colony, mix one-part white cane sugar with one-part warm water and stir until dissolved. There is no need to boil the sugar water. Make sure to buy several bags of white cane sugar before hiving your new bees. Never feed bees brown sugar or honey from unknown sources.

A package of bees should be installed in a hive with only one deep box. Do not add extra boxes until the colony population grows. Bees do not need more space than they can patrol and protect from pests. This small colony will have too much room if you add several boxes at first. When the bees have filled 80 percent of the frames inside the hive body, it is time to add the next box. This will take several weeks or more.

Don't expect to harvest honey during the hive's first year. For most beekeepers, it is usually the second year before the bees have excess honey. However, foraging conditions do vary. If you are lucky enough to live somewhere with a long bloom season, you make get a bit of honey sooner. Be patient.

Most beekeeping equipment is purchased unassembled. Putting your equipment together is a good exercise for a new beekeeper. Assemble your equipment using nails and wood glue. Outer boxes are exposed to weather and frames take a lot of abuse over the years. Failure to use glue can result in frames coming apart at very inconvenient times.

Do you have to paint your hive? No, the bees don't care. However, the wooden parts of your hive will last much longer if they are painted. White is a traditional color for beehives but you can use any color of exterior latex paint. Express yourself with bright eye-catching colors and patterns, or blend your hives into the landscape with tans, light browns, or greens.

Protective Beekeeper Clothing

Now that your beehive is prepared, what about you? Yes, you need bee stuff too. Rushing out to buy an expensive bee suit is not a requirement of beekeeping. A pair of long pants tucked into your socks and a long sleeve light colored shirt is a good start. Now, add a beekeeper's hat and veil to complete your ensemble. A bee sting in the eye could result in blindness. There is no reason to take that risk. You will find many styles of hat/veil combos in bee supply catalogs. Quality varies so make sure you are getting good quality if you are paying a premium price.

Most new beekeepers choose to use a beekeeping jacket or a full bee suit. Many of us beekeepers with years of experience wear jackets or full suits, too! Your beekeeping wear should be comfortable and loose. Remember, you will have to twist and bend while wearing it. Suits and jackets are available in different styles, materials, and colors. Light colors are best but your clothing does not have to be white. Just try to avoid dark colors as you do not want to look like a bear.

Beekeeping gloves are a subject of some controversy in beekeeping circles. While providing protection from stings, thick gloves can cause clumsiness during hive inspections. Thin beekeeping gloves made of goatskin instead of leather are a good alternative.

Sadly, some beekeepers might tell you that you should not wear protective gear. But do not be too embarrassed to wear a full beekeeping suit or jacket and gloves. This is your beekeeping journey. Please yourself and do not let others dictate what you should wear. You want to be as calm and confident as possible. The bees can sense tension. Wear your beekeeping suit and be a confident new beekeeper. Tell those who criticize you to buzz off!

Beekeeper Tools

Do you like to buy gadgets? If so, you are in luck. Beekeeping catalogs contain essential tools and hundreds of gadgets. You may feel overwhelmed at the selection of tools. Each one promises to make your beekeeping life easier, but do you really need them all? No. Purchasing beekeeping gadgets is a lot of fun. But some are useful and others are a waste of money. Start with the basics and save the specialty tools for later. Beekeepers disagree about many things, but all beekeepers agree on two tools needed by every beekeeper: the beekeeper's smoker and a good hive too (flat metal tool).

Bee Smoker

The beekeeper's smoker is an iconic representation of beekeeping. A metal can with a handle, bellows, and a spouted top for smoke release will become one of your best friends. It will last for many years so don't begrudge the money spent on a smoker. Choose one with a safety guard around the fire chamber. Trust me, if you do not, you will grab it someday and burn yourself. Working with your smoker will be easier if you choose the right material for fuel. Dry pine needles, wood pellets, untreated burlap, and commercial smoker fuel are common choices.

Learning to light and use a smoker can take some practice. Do not feel discouraged if it takes time to perfect the technique. I have been using bee smokers for years and still have trouble sometimes. (Can't get more than a puff of smoke in the bee yard—yet it sits puffing away under the shed when I am finished!) Use a lighter or match to ignite a small piece of crumpled newspaper or pulp egg carton as a good starter. Toss it into the burning chamber and gently puff the bellows. This will add oxygen to the flame. Add a few more small pieces of the starter material while continuing gentle puffs of the bellows. When you have a reasonable little flame going, it is time to add a bit of your smoker fuel. Push the material down into the chamber with your hive tool. Be careful not to burn yourself. Do not pack the fuel down tightly. Puff the bellows a while longer. If the smoker is producing a good white smoke, it is time to add more smoker fuel and pack it firmly in the burning chamber. Close the top and you should have cool white smoke.

Bee smoker on hive

Do not use accelerants (toxic fuels) to light your smoker as this can be toxic to the bees and you. The smoker should produce cool white smoke, not dark hot smoke. Puff a small amount at the hive entrance and under the top lid. Wait a few minutes before beginning your inspection. A small amount of cool smoke disrupts the bees' communication system. This slows the alarm message sent out by guard bees when you open a hive. The bees may not like the smoke but there is no lasting harm.

Be careful with your smoker. Beekeepers have set their cars on fire with unattended smokers. And at least one beekeeper (me) has been pulled over by a cop because of a smoker breathing smoke in the back of the truck. It was safe but the policeman didn't know it at first glance.

Hive Tools

Hive tools are a very basic tool. How many hive tools do I have? A lot. Can I find them right now? Well, maybe I can find one. Hive tools are notorious for being easy to lose. You can never have too many but you need at least one. I use my hive tool to separate hive boxes that are stuck together, to push fuel down in my smoker, and to pry frames out of a hive. Frames must be removed from the hive for inspection but bees often glue them down with propolis. (Remember the propolis we talked about earlier on page 21?) It is also called "bee glue." Honeybees collect plant resins from tree buds that is mixed with saliva to form a sticky substance called propolis. Bees use propolis to polish and stick down everything in the hive. Propolis has antibacterial properties and is a component in various medicines and tinctures. For the beekeeper, propolis is an aggravation. A good hive tool eases removal of stuck frames, and any southern beekeeper is familiar with the delight of smashing small hive beetles with a hive tool.

Chapter 6

Your Bee Yard or Apiary

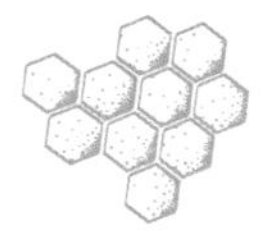

Why Moving Hives Is a Big Deal

Every beekeeper needs to move a hive from time to time, but moving hives to a new location is a big deal to the honeybees who live there, so think carefully when choosing a new hive location. Avoid putting your hives in temporary locations with plans to move them later. Moving a hive to different locations around your backyard is especially hard on your bees. Why does moving a beehive ten feet cause a bigger problem than moving it one mile or more? It's because of how honeybees find their way home.

Honeybees have a special way of navigation. In addition to visual cues, they have an ability that sounds much like a superpower. They learn the location of their hive by using the magnetic fields of the earth and the position of the sun. If you move a beehive ten feet to the left or right it causes a problem for the bees because many foragers will instinctively return to the old hive entrance and may not be able to find the hive in its new location that is only a few feet away. Some bees will be lost. However, when a beehive is moved one to two miles away, bees emerging from the hive will sense the move. They will reprogram the new location in their brains, and foragers will be able to return to the new hive entrance. If you absolutely must relocate your hive, it is better to move the hive a mile or more away for a few days, then return the hive to your desired location in your bee yard. This movement plan makes reorienting easier for the bees.

Perhaps you do not have time to move heavy hives. Or, you may not have a friend willing to allow a beehive in their yard for a couple of days. So, here is the work-around. You can move a hive to a new location nearby the hive in small increments at a time. Move it two feet, leave it for two to three days, then move it two more feet. This is time-consuming and requires many trips to the hive, but moving the hive in small amounts allows bees to follow the movement.

One last word of advice: do not place your beehive close to your back door just so you can enjoy seeing them come and go. The temperament of a beehive can change quickly. You may enjoy having the bees close to your dwelling but they may not feel the same about you. A change in hive temperament could lead to a danger situation for any humans or pets in the area. As we consider how bees orient, it is easy to understand the importance of hive placement.

Choosing a Good Hive Location

Now that we know a lot more about how bees find their way home, what other factors should we consider when choosing a good location? How do you find the best spot for your bees? Honeybee colonies thrive in many locations. On a rooftop in New York City, in a suburban backyard, or on a farm in Wisconsin, bees can be kept almost anywhere.

Should you choose a sunny spot for your bee yard? In the past, beekeepers placed hives in morning sun and afternoon shade. This is still a good practice in areas that experience brutal summer heat, but placing colonies in full sun aids in preventing large beetle populations. Beetles easily reproduce in damp, sandy soil conditions. In regions affected by small hive beetles, place your hive in a dry location with more sun than shade. Small hive beetles are difficult for honeybees to control due to the hard beetle shell. A savvy beekeeper does everything possible to limit beetle numbers. Good hive placement is one thing you can do to help the bees deal with small hive beetles.

Do you need a stand for your beehive? While not a requirement, having hives on a stand makes beekeeping easier. You can build wooden hive stands, purchase metal stands, or set the hives on a few cement blocks. Raising the hive off the ground makes it easier to work. Your back will thank you for years to come. It also protects the hive entrance from skunks, and protects the wooden bottom from rot. A hive stand must be strong and sturdy. It

Beehives on a stand

needs to support a beehive that can weigh over 400 pounds by year-end. Finding your bees on the ground because you set up a wimpy stand is not a pleasant experience.

Providing Bees Access to Water

Honeybee colonies need water, and they do a great job of seeking out water sources. When planning hive placement, it is good to give some thought to where your bees may look for water. Bees can fly, so water does not need to be right next to the hive. Any clean water source within a quarter mile is okay.

Bees use water to thin honey for food and to cool the hive. Bees collect water year-round on warm days but they harvest much more during the summer heat. If you do not have a natural water source nearby, you should provide water. I have always been an avid plant lover. Before becoming a beekeeper, I built several small water gardens on my property. Now, my water gardens

Water source for bees

provide clean drinking water for my honeybees. The water source needs to have a shallow drinking area with pebbles or sand, otherwise some bees will drown. Provide a water source large enough to last for several days, even in hot weather. You do not want to have to fill it every day. When the weather is hot, bees can deplete a small water feature quickly. If you or your neighbor have a swimming pool, providing water becomes even more important. The water source should be closer to the beehives than your neighbors swimming pool. Once bees imprint on a water source, it is difficult to keep them away. It is a good plan to have a water source in place before your bees arrive.

Winds, Floods, and Mud

Protecting your hives from extreme weather is important for several reasons. Do you live in a region with strong cold winds? Consider planting tall shrubs on the windward side of your hives to provide a windbreak. Cold winds blowing in the front door make it very difficult for bees to stay warm. Bees do not heat the whole interior of the hive but they do keep the cluster warm. Drafty conditions make that difficult.

Keeping hives close to a beautiful stream sounds like a good idea, but weather is changeable. Placing your hives in a flood zone is asking for trouble. Honeybees can fly but they are not great swimmers. Soggy beehives floating down the river are a sad sight. It is not necessary to place your colonies very close to water; they can fly. (I keep saying that, don't I?) Placing bee colonies in damp low-lying areas is also bad for their health. Too much humidity encourages disease.

A beekeeper needs to be able to access hives at all times. Can you get to your apiary in all types of weather? Regular hive inspections need to be completed, especially during the warmer months. Beehives may need relocation due to a predator attack or human thievery. Some colonies need to be fed and that requires transporting sugar water. Honey harvesting is heavy work as well. Several bee management factors may require getting a vehicle, ATV, or wheelbarrow to the bee yard. Keep this in mind when deciding where your apiary should be located.

Make an equipment plan for "pre-bee" arrival. Choose the best location, prepare your hive stand, and have everything ready a couple of weeks early. Bees are very sensitive to odors. It is not a good idea to place bees into a freshly painted hive. The scent alone may cause your bees to leave. Also, don't wait until the last minute to order equipment. Spring is a busy season and suppliers often run out of stock on popular items. Have all necessary equipment on hand a couple of weeks before your bees arrive. All hive components must be ready days before installing bees. Good planning makes the process easier, but if you forget something that you need, it will be okay. Trust me.

Chapter 7
The Bees Arrive

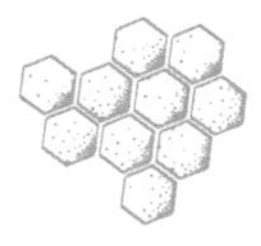

The vast majority of new beekeepers begin with package bees. Many suppliers offer bees for sale in late winter for spring delivery. The three-pound size is most common and contains roughly 10,000 bees. Package arrival season peaks in early spring and most bees are delivered in April and May. If you wait until spring to order bees, you may be out of luck. Order early.

Package Bees

The day you have waited for is here; your first package of honeybees has arrived. You feel the excitement and adrenaline building. You can do this! Can't you? Yes, you can. If your bee package is arriving through the mail, be prepared for an early phone call. The local post office will probably be anxious to hand off your buzzing darlings. Traveling in either a plastic cage or one of wood and wire, the bees are not too happy about being there either. On the way home with the package, keep them cool. This assumes that you live in an area with moderate temperatures. The bees inside the transportation box are not able to cool themselves as they would inside a beehive. With temperatures in the 50s, the package can sit in the back of the truck on the way home. I would push them up near the cab out of the wind. It is more common for beekeepers to overheat their packages than to let them get too cold.

Perhaps you will travel a couple of hours to pick up your new bees from a bee supply. If so, please go straight home with your bees. Don't stop to have lunch and leave them in the hot car. If you must make a pit stop, park in the shade, or leave the air conditioning on low. Upon arriving home, your package can sit in a shady spot while you get ready to hive them. Some resources suggest waiting until late afternoon to install packages. That is a good idea. For myself, I install packages as soon as possible. I feel it is important to get them back into a more natural environment.

Before installing your bees in their new home, give them a little snack. Mix up some sugar water and put it in a spray bottle. The mixture should be about half white cane sugar and half water. Shake well. Lightly spray the screen sides of the bee package. This gives the bees a snack and helps reduce the number of bees that fly out of the package later. The bees in your package are rather lost and confused but are ready to get out!

The queen bee will be inside a small wooden/wire cage hanging inside the package. She has several workers with her. These queen attendants

Queen cage with candy

came from the same hive as the queen. They will feed and groom her. Depending on the bee supplier, your queen could also be in a plastic cage. The queen cage will have some white candy in one end. This candy feeds the attendant bees and allows the queen to be slowly released over a few days. Don't immediately turn your queen loose. The bees in the package did not come from the same hive as the queen. They do not know her and may kill her if she is released too soon. In most cases your queen bee will be okay. However, always check to see that the queen in the cage is alive. Your queen may or may not be marked with a dot of paint. This varies from one supplier to another. If you are given the choice when ordering bees, always choose to receive a marked queen.

If you receive a package of bees with a dead queen in the cage, don't panic. The supplier may have accidentally shaken another queen in the package! A loose queen will seek out and kill the caged queen. Most suppliers will replace a dead queen if you notify them right away. It may be necessary for you to return the dead queen in the cage, so hold onto it. The supplier may instruct you to wait a day or two and check for a loose queen in the mass of bees. It is still important to let the supplier know right away that you may have a problem. Quality bee businesses want you to succeed and will do what they can to help.

Package Installation

There are several ways to install a package of bees into a hive. My favorite method is quick and easy but all methods work well.

Method 1: Shake Them In

Wooden bee packages have a flat piece of wood stapled over the top opening. Use your hive tool to pry this off. Don't worry, bees will not come flooding out. You will see a large circular hole and a silver can. The can contains syrup to feed the bees during transit. It fits inside the access hole. The can has a tiny hole on the bottom allowing the bees to drink. Larger holes would allow the syrup to be shaken out during transport. Rarely, but it does happen, the supplier will not get a good hole in the can. We call this a "dry package." The bees will be hungry and angry, but a gentle sugar-water spray on the screen often helps to calm them down. A little is enough, so make sure not to overdo it.

After spraying the screen sides of the transportation box with sugar water, gently bump the package on the ground (not too hard). You want to bump the package just hard enough for most of the bees to fall to the bottom of the box.

Getting the syrup can out of the package can be aggravating. If you have needle-nose pliers, you can grasp the can and slowly lift it out. Bees will be hanging from the bottom; it's okay. If you do not have pliers, pick up the bee package and tilt it slightly to the side causing the can to drop out enough for you to grab. This is quite scary for new beekeepers because it feels like you are trying to pour out your bees. Be gentle. They will be okay.

Remove the syrup can and set it aside. A plastic tape sticks out of the circular opening. At the other end of that tape, your queen bee is inside her queen cage. Grab the plastic tape holding the queen cage and pull the tape free from the top of the box. You can place the flat piece of wood back over the hole to keep most of the bees inside. You will have some bees flying around. That is okay.

Having removed five frames (with foundation) from the ten-frame beehive, you have open space to work with in the hive. If tilting the package and removing the can is too much for you, you can simply remove the screen from one side of the shipping box and pour the bees into the hive. If you do that, it is even more important to get that queen cage installed quickly so the bees find her in their new home. A better plan is to place your queen cage in the new hive before releasing the majority of the package bees. Each end of the wooden queen cage will have a cork preventing bee escape. On one end, you will see a mass of white candy; this is the end from which you will remove the cork. The bees will slowly eat through the candy plug. This will take several days, giving the bees time to grow accustomed to the queen's pheromones. Do not open the cork on the non-candy end! Doing so may result in your queen being killed. The bees don't know her yet. It takes time. These instructions apply to a wooden queen cage. If your cage is plastic, you will not have corks. But you may have a plastic cap on one end of the queen cage.

After removing the cork plug from the candy end of the queen cage, hang the queen cage between two frames in the hive. You can use a thumbtack to secure the plastic tape to one of the top bars. Make sure the wire side of the queen cage is turned to allow bees access to the queen. Your package bees will feed the queen through the wire and get to know her. At this point, your queen cage is installed. Do not poke a hole in the queen candy plug. Some resources will tell you to do this but it is usually not necessary. If you check the colony after seven days and the queen is still not out of her cage, just release her. Many nervous new beekeepers have skewered their queen attempting to put a hole in the candy.

Now it's time to carefully add the rest of the bees. Pour the package bees into the hive's empty space. Shake most of them out of the package. You won't get them all out. Carefully replace the remaining five frames in the hive (for a total of ten) and push the frames together. The package bees will find the queen cage and cluster around it. Place the mostly empty package in front of the hive. The remaining bees should join their sisters inside the hive before dark.

Method 2: Slow Release

This method differs only in how the bulk of the package bees are put into the hive. Remove five frames from your new hive. Hang your queen cage between two of the remaining frames in the hive. Use a small wire to hang the queen cage in the hive. Or, you can use a thumbtack and the plastic strip on the cage. Hang the queen cage with the candy end pointing up or to the side. Remove the cork from candy end only. You do not want the candy end facing down because if one of the workers inside dies, it may prevent the queen from escaping the cage. Remove the syrup can and place the open bee package into the empty spot in the hive. If you use this method, gently shake a few bees out over the queen cage. Close the hive. Over the next few hours, bees will leave the open package and go to the queen. The next day you can remove the empty package and replace the five frames you removed earlier. This is a good way to install package bees but it is not my favorite. I dislike having to disturb the bees on the next day to retrieve the empty package.

And sometimes the bees don't leave the package, leaving the queen to possibly die on a cool night.

You did it. Your queen bee is hanging inside the hive protected by her queen cage. You removed the cork from the candy end of the queen cage (if wood) and not the wrong end. The bulk of the bees are inside the hive and will soon get down to "bee business." You are finished, right? No, not quite.

Feeding Your New Bees

So much effort has gone into preparing for your new bees. It would be a shame to let them starve now. Any new colony has a lot of work to do before winter. A package of bees is not a normal bee swarm. They were in a hive going about their "bee business" when someone shook them out of the box into the package. They were not prepared to start a new home and did not bring any honey stores with them. Natural bee swarms actually fill their honey stomachs before leaving the hive. (A snack for the road, you know.) This is because bees will starve in a few hours without food. The success of your bees will depend a lot on the care you give them for the next two to three months. A new bee colony needs to be fed.

Growing colonies need a 1:1 sugar-water mixture (equal amounts of white cane sugar and water). Failure to feed new packages properly is one of the main reasons for hive failure in my area. Continue to offer syrup to new colonies until they have completed building honeycomb on all frames and storing some honey in two hive boxes. (For me, that is one deep and one shallow.)

The colony may slow the intake of food during the spring bloom. Having feeders will allow them to feed at night, on rainy days, and days that are too windy for flight. When preparing to feed your new bee packages, think in terms of gallons, not quarts.

You do not have to buy a bee feeder if you have an extra deep hive body. An old standby for feeding bees is using glass quart jars with a few small holes punched in the lid. This jar of sugar water is set upside down on the top bars of the frames. An empty hive box is added to enclose the jars and allow the telescoping top to close the hive. As the colony grows, this box must be removed and the hive given new space. Otherwise, they will build comb in and around the jars; that is a hot mess.

A beekeeper who feeds a new hive for only a couple of weeks may find a surprise in the September hive. No food stores for winter and frames that do not have honeycomb drawn out are common. The new colony full of potential in the spring now faces starvation with cold weather making feeding difficult or impossible.

The moral of this story is feed your bees. If they lose interest in the sugar syrup, remove it for a few weeks. Then offer a small amount again. If they take it quickly, you will know it is time to start feeding again. How much honey does your colony need for winter? That will depend on your location or climate. In my region, my bees do well with a shallow super full of honey in addition to the food stores in their deep box. Your bees may require two deep boxes. Ask local beekeepers what works in your region.

The Ups and Downs of New Hive Management

Most package queens are released within three to five days of installing the package. The colony has acclimated to her pheromones and accepts her as their new queen. In a few days, she begins to lay eggs in the newly constructed comb. The life cycle of a worker bee from egg to adult is twenty-one days. During this time, the population of your new hive will drop. Bees die of old age every day. However, once we reach between twenty-one and twenty-eight days, the population should grow very fast.

Check Your Queen Status

The first visit inside the new hive will be to remove the queen cage, four to seven days after installation. Carefully check to see if she is released from the queen cage. If so, remove the queen cage but don't go looking for her on every frame. Don't stress your new colony. This is not the best time to look at each bee on each frame in the hive. Your bees are still settling down in their new home. Pull the queen cage out, push the frames together, and close the hive. Be sure to gently push your frames together, leaving any extra room divided on each side of the box. Sometimes you need to wedge frames apart a bit to fit the queen cage inside. But we don't want to leave it that way long-term.

What happens if your queen and her friends are still in the cage? Occasionally the queen candy (white plug) is too dry and it takes the bees longer to release the queen. If your queen is not released by day four, recheck again in a couple of days. Around day seven, you can release her. Don't get in too big a hurry to release your queen. This applies to both new colonies and those being re-queened. The introduction phase is important and should not be rushed.

Things normally happen in a good way and your new package grows into a strong hive. But occasionally, things go wrong. The queen may die before escaping the queen cage. The bee colony may refuse to accept her and kill her upon emergence. She may not be well-mated and be incapable of producing a good compact brood pattern. You need to know what is going on in the hive. But you do not want to bother your bees too much while they are getting used to their new home. Each year, I talk to new beekeepers who go inside their hives every day. Some of those bee colonies simply fly away or abscond. Inspect your hive; you have a lot to learn, but don't disturb your bees every day.

Check the queen status of your hive one week after installing the package. You are looking for tiny eggs or white "grub-like" larvae at this point. If you see eggs/larvae, close the hive; leave them alone. Recheck in another week; you should see larvae and capped brood by now. Capped brood refers to young bees in the pupal phase. Workers have sealed the cells to allow the young bee to develop. If you see the queen but no brood, recheck in another four days. If you see no brood day days after install, your colony may have a queen problem. Contact your package supplier for advice.

New honeybee colonies have a lot to do during their first season. It is a tenuous time for the colony. Comb must be built, food stored for winter, and young raised all at the same time. For new hives, perform hive inspections weekly until the colony is about three to four weeks old. By this point, you should have an idea of the quality of their queen. But avoid opening your hive too frequently.

Bee eggs

Great brood pattern

Beekeepers often mark the thorax of queens with special paint markers. This makes finding your queen bee easier. This is especially true during summer in a colony of 60,000 bees. Marking with different colors also helps identify the age of the queen bee. You can mark your queen with any color but most beekeepers use the international marking system: white for years ending in one or six; yellow for years ending in two or seven; red for years ending in three or eight; green for years ending in four or nine; and blue for years ending in five or zero.

Chapter 8

Finishing Out Your First Season

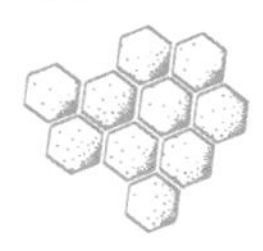

The First Summer

Throughout the summer, beekeepers monitor their colonies closely. Ensuring the hives will be healthy and filled with honey for winter is an important concern. Most beekeepers choose to feed new colonies. This gives the bees their best chance of survival. If you live in a warm climate with year-round blooming flowers, feeding is not as critical. However, feeding does aid in the growth of new colonies. And food stores are a concern for the beekeeper until the bees have enough food for winter. In southern regions, this is one deep box and one shallow super box at minimum. Do not try to overwinter a bee colony with only one deep box unless you live in a tropical region. Each frame in the hive should have comb built out and be in use. The colony should have a good population of adults and ample stored honey and pollen before winter arrival.

Monthly hive inspections should verify the presence of a laying queen. Finding the queen can be difficult in a populous hive. However, seeing a good pattern of normal worker brood indicates a good queen status. You should always see some worker brood during summer months.

Varroa mite infestations are always a concern. Do not assume that your bees are okay because you don't see mites. Develop a mite control plan and put it into practice. Yes, even first year colonies can crash from problems associated with mites.

Hive Inspection Techniques

Looking inside the hive can be very educational. It is one of the most useful types of training for new beekeepers. No book can prepare you for the wonder and mystery of seeing bees working on the comb. I still find it remarkable to be able to lift a frame of bees from the hive and observe their behavior. Most of the honeybees continue their tasks and ignore me. Honeybees are so obliging.

As intriguing as it is to observe the bees, any inspection causes a disruption to the colony. We must remember that it is not natural for a hive to be open to light. Or, to have a giant in a white bee suit removing parts of the bees' home. All hive inspections should have a goal and be as brief as possible. You can watch a million videos but the live experience is much different when you are looking inside your own beehive in your backyard.

Ideally, inspect your hive on a warm, fair day when the older foragers are out working. Late morning or early afternoon is an ideal inspection time. But don't delay needed inspections waiting for the perfect day. Wear your protective gear and have the smoker lit. We want cool white smoke. Move slowly around the hive and avoid banging hive parts together; bees react to those types of vibrations. Lift the frames gently and avoid crushing bees as much as possible. Stand beside or behind the hive so you are out of the flight path.

Following a routine inspection schedule is a good management strategy. However, some hives will have issues that need more beekeeper attention. If you notice possible problems such as less flight activity at the front, evidence of robbing bees, or bees that were calm previously becoming more defensive, a quick check is advisable. A hive with a problem may need a weekly inspection for a while and then be okay on a monthly schedule.

Monthly checks are a good rule of thumb for established colonies except during swarm season. During spring when there is a stronger chance of swarming, inspect strong overwintered colonies every two weeks. Does the colony seem very crowded? Watch for queen cell development, especially on the bottoms of frames. If you see signs of swarm preparation, you will need to decide how you want to handle it. Let the bees swarm and hope to catch the swarm or split the colony into two smaller colonies before the bees do it themselves.

Things go wrong sometimes. A hive inspection can unveil many problems. If something goes wrong that the bees can't fix, you need to know. This is true whether it is your first year as a beekeeper or your tenth.

Key Things to Look for Inside a Beehive

As a beekeeper, you want to see normal growth inside the hive. Learn to recognize the various stages of brood. Do you see eggs, larvae, or capped brood? You do not have to see the queen but you must see evidence of a laying queen. What does the brood pattern look like? Do you see brood of similar ages close together with only a few empty cells? Spotty patches of brood spread here and there may signal a problem with your queen.

Checking for Brood

What does your brood pattern look like? This is a common question an experienced beekeeper asks when a new beekeeper expresses concern for a hive. Beekeepers spend a lot of time thinking about brood because a constant supply of new bees is vital to a healthy colony. The brood nest refers to the area where young bees are being raised. Brood pattern refers to the different types of brood (egg, larvae, capped brood) and how they are concentrated in the brood nest region.

A queen bee can lay thousands of eggs per day. Tiny white bee eggs are hard to see but look like small white threads in the bottom of a honeycomb cell. Each cell should only have one egg. In just a few days, those eggs become larvae. Larvae are white grub-like baby bees; we call them "milk brood." This phrase is used due to the milky white brood food seen in the bottom of the cell. Larvae actually

Milk brood in cells

Worker bees

lay in a pool of food. If you see small larvae in your hive, it is a good sign. Your colony is probably "queen right." This means a queen has been in that area of the hive recently. When larvae reach the pupal growth stage (around day nine or ten) they no longer need to be fed. Adult workers cap the cells with old wax that often looks tan or brown. At this stage, they are called capped brood. Capped worker brood is easier to find than eggs.

Beekeepers want a tight pattern of brood grouped together with few empty cells. It is okay and desirable to have a small number of cells empty. But it is not a good sign to see patches of brood scattered here and there. If you only see milk brood, you will want to recheck in a week to confirm that it has progressed to the capped stage. Occasionally, a queen is not properly mated and she will look great but not be able to produce a good pattern of worker brood.

Finding Drone Brood during a Hive Inspection

As your beekeeping experience grows you will learn to recognize the difference between worker brood and drone brood. Capped (female) worker brood cells protrude very little above the comb surface. Drone (male) brood cells are larger than worker brood cells and the brood larvae are longer. The cappings of drone cells protrude noticeably from the comb surface. They are sometimes described as "bullet" shaped.

New beekeepers often confuse drone brood for queen cells. However, drone cells do not hand down the face of the comb like queen cells. It is natural to have drone cells in your colony during the warm season. They are usually on the outer edges of the brood nest. Drones fulfill an important role in the honeybee colony. However, if you have nothing but drone brood, you have a problem! Perhaps your hive has lost its queen or the queen is no good.

A mated queen stores sperm inside her body. When her supply of sperm runs out, she can only lay unfertilized eggs, or drones. The sign of a failed queen is a large amount of drone brood with little or no regular capped worker brood. The colony may recognize that the colony is failing and try to replace her by making a new queen. If not, the beekeeper must get a new queen for the colony. Finding worker brood is always a good thing. Seeing drone brood is a good thing too, but all drone brood with no worker brood is not a good sign for any colony.

Laying Workers

A bee colony may become queenless and fail to raise a new queen. Or, a beekeeper may add a new queen that is not accepted by the colony. If they don't want her, they will kill her. Slow introduction with a queen cage reduces the likelihood of this, but it can happen. When a honeybee colony is without a queen or brood for several weeks, some of the workers will start to lay eggs. We call these "laying workers" and they are a problem. Because they cannot mate, these workers can only produce drones. Finding drone brood scattered everywhere and seeing many cells with multiple eggs in each cell are indicators of a laying worker problem.

The laying worker bees fly outside the hive and forage. They cannot be identified on sight because they look the same as the other workers. The colony is doomed without beekeeper intervention. There are several techniques that supposedly take care of the laying worker problem. The easiest is to combine the laying worker colony with a queen-right hive. Once the laying workers sense the presence of a queen and normal brood, they will cease laying eggs.

What Is Burr Comb?

Building comb (also called "drawing comb" or drawn comb) is hard work for our bees. They need to consume a lot of nectar to make beeswax. During hive inspections, watch for strange comb. As you look down through the frames, does something look weird? You may find some awkward comb built around the queen cage. This is normal and you can remove it and push the frames back together. Do you see "cross comb" where two frames are connected together by comb? We call this "oddly placed" comb, or burr comb.

Burr comb may also appear on the tops of frames indicating that the wooden ware is not cut precisely enough to preserve the proper spacing desired by the bees. This issue is a bigger problem for the beekeeper than the bees. You can scrape the burr comb off, but if your bee space is wrong, the bees will just build it back. However, small hive beetles can hide in burr comb, so most beekeepers scrape it off the top bars.

Too much burr comb between the frames is a bigger problem than extra comb on top. Remove any comb that is not being constructed where it should and in the proper manner. Wonky comb makes future hive inspections more difficult. In a hive without proper frame spacing (or with new plastic foundation), bees will sometimes build another sheet of comb outside of the frame. The queen will lay there, and it is a mess, so remove any unauthorized comb placement. The sooner it is removed the better. The queen bee will lay eggs in this misplaced comb.

After any hive inspection, always make sure the frames are pushed together before closing the hive. Extra space can be divided on each side of the hive box. Unless you are using frame spacers, you do not want to try to evenly space the frames out. They are designed to be pushed closely together.

Hive Inspection Tips to Remember

- You don't have to see the queen every time you inspect; look for eggs and young larvae
- Look for a good brood pattern with few empty cells amongst the brood
- Use cool white smoke a few minutes before opening the hive
- Don't spend too much time with the hive open
- Look for stored pollen in cells (in shades of yellow, gray, or red)
- Do you see any honey? Make sure your bees are not starving
- Look for signs of mites or other pest problems

Putting the Bees to Bed for Winter

Your bees should have enough honey stored for winter, well before cold weather arrives. Once temperatures drop below 57 °F, feeding bees becomes more difficult. The honeybees will cluster together for warmth and not leave to reach the feeder. Evaluate your beehives several weeks before your last expected frost date. If you need to feed them, do so.

Some beekeepers are against feeding honeybees. However, when there is a drought, I feed my livestock. Feed your bees if needed. Before cold arrives, you should have implemented your varroa management plan. We want varroa mite numbers to be under control by the end of July. This gives the colonies several months to raise healthy winter bees.

As the days grow shorter, most queens slow down egg laying. Seeing less brood in the hive at this time of year is normal. If your colony has enough food stored (how much needed depends on your location), has a good healthy population, and no pest or disease problems, they are ready for winter. You cannot wait until the cold weather of fall to feed your bees or treat for varroa. By October, it will most likely be too late. Start preparing your bees for winter in August or before!

The Winter Cluster

Honeybees are able to survive the cold winter temperatures because of two management techniques. Bees store a large amount of surplus honey and they can generate heat! When temperatures drop below 57°F, the bees cluster close together inside the hive. This ball of bees consists of thousands of individuals. Honeybees are able to "unhook" their wings and vibrate wing muscles to create heat. However, this cluster must stay in constant contact with food/honey.

Some bee colonies die every winter. It is normal to feel disappointment when this happens. Even beekeepers with years of experience lose hives. There are too many variables involved in the management of these insects to ensure zero losses. The beekeeper's job is to manage colonies in the best way possible based on what we know of their needs. Do everything you can to help your bee colonies stay strong and healthy. Good hive locations, proper inspections, varroa mite control, and proper feeding will all help increase your chance of beekeeping success. When you lose hives, and you will, do not become too discouraged. Try to evaluate why the colony failed and how you might avoid similar losses in the future.

Your Honey Harvest

While it is possible to have a honey harvest the first year, this is not the norm in many regions. New colonies have a lot of work to do in that first season. Construction of honeycomb, building population, and building up winter food stores all require the efforts of thousands of worker bees. Most locations do not have a season long enough for excess honey production by new colonies. In fact, many colonies fail in the first year unless the beekeeper feeds them well.

Don't be too impatient for a honey harvest. A healthy, productive colony can produce a larger honey crop for you next year. Avoid robbing the bees because you can't wait and leaving them to starve. Before you know it, you will have a golden harvest of your own.

Extracting honey

Chapter 9

Swarming

The bee swarm is one of the most fascinating sights in the insect world. What is a swarm? A bee swarm is a large mass of honeybees that leave their hive to create a new home somewhere else. Approximately half the bees in the colony and a queen bee leave the mother hive. However, some of the bees remain at the original hive. They will produce a new queen and regrow the population to continue.

Swarming is reproduction on the colony level for honeybees. They make more bee colonies by growing a large family that then splits into two hives. This results in two smaller colonies complete with a queen, workers, young, etc. Both colonies must work hard to establish themselves before winter. This is especially true for the swarm of bees that must build a new home from scratch.

Swarm prevention is a hot topic in beekeeping circles. Many beekeeping books give advice on how a beekeeper can prevent swarming. The truth is that sometimes you just can't stop this natural impulse. A swarming hive produces less of a honey crop so beekeepers often strive to reduce swarming. Hives created with package bees are less likely to swarm the first year. In fact, some resources will tell you that first year hives will not swarm. Don't believe it. They certainly can and do sometimes.

There are several swarming triggers. These are the actual cues that initiate swarm preparations inside the colony. Though it may seem to be the case, bees do not swarm on an impulse. Plans begin weeks before the swarm leaves. A healthy bee colony grows strong quickly during spring. Nectar-producing flowers begin to bloom and fresh pollen becomes available. With the influx of food, the queen increases egg production. Before long, she may have trouble finding cells that are not in use. As more and more bees occupy the beehive, queen pheromones are diluted in the colony. Both lack of space and large populations can signal the beginnings of a bee swarm. In addition to crowding or congestion, a colony with an older queen is more likely to swarm. This is one reason that beekeepers like to keep a young queen in their hives.

A honeybee colony can "throw a swarm" at any time during the warm season. Most swarming events will occur in the spring as this is the natural time of growth. However, it is not unusual to find swarming colonies during summer or even fall. A colony in "swarm mode" makes preparations to produce a new queen bee. The old queen usually leaves with the swarm. Several queen cells, or swarm cells are constructed. Young female larvae are fed a rich diet to promote queen development. Scout bees are workers who fly out looking for resources needed by the colony. In pre-swarm mode, the scouts are looking for possible nest sites. When all preparations are completed, the bee swarm is ready to form and leave the hive.

On a warm, fair day usually between 10:00 a.m. and 2:00 p.m. the honeybee colony becomes very active. A strange roar can be heard inside the beehive. Just before leaving, worker bees fill their honey stomachs with honey. This will keep them alive for a couple of days until a food source is found. The queen bee and up to half the workforce leave the mother hive. A swarm includes a queen, thousands of workers, and drones as well. The swarm finds a temporary resting place somewhere near the hive. A tree or large bush is a common transition spot. Within a few hours, the bees take flight again, headed to the new nest site. Inside the mother colony, the remaining bees carefully tend capped queen cells. In a couple of days, a new queen will emerge. The strongest virgin queen will kill her rival queens and go on to mate and become the new queen bee.

If you find a swarm hanging in a tree or bush, they are rather easy to catch. There is no need for fear as swarms are not aggressive. However, if threatened, they will defend themselves. Always keep pets and children a respectful distance from a swarm of bees.

Swarm Management Tips

- Replace older queens before the bees decide to do so
- Give a growing colony space; when one box is 80 percent full, add another hive
- If you see queen cells in a strong colony, split the hive before they do
- Make sure the queen has room to lay
- Always keep extra equipment on hand to catch swarms
- Place swarm traps in strategic places around your area

Chapter 10

Disease and Pest Control in Beehives

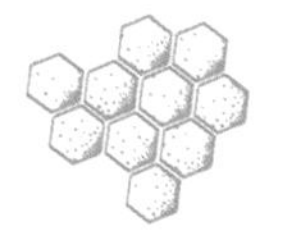

Varroa Mites

Varroa mites feeding on worker bee

The arrival of varroa mites signaled a changed in American beekeeping. Hive management procedures now must always include a varroa mite control plan. This external pest feeds on adult bees and brood. The mites' feeding activity weakens bees and causes the introduction and spread of viruses. Most colonies that are not treated for varroa will die within two years. Some sources will say that package bees do not need to be treated the first year. Don't believe it. Always monitor mite levels in your hives.

Developing Mite-Resistant Bees

Researchers are working on the development of new lines of honeybees. Through selective bee breeding, beekeepers hope to see bees with behaviors that will allow them to better deal with varroa mite infestations. The term "hygienic" is often used when discussing bee breeding. In time, this is probably our best hope for beekeeping without chemicals.

Honeybees in a hive with hygienic tendencies will vigorously groom themselves. In the grooming process, mites fall off before they can enter a brood cell to feed and reproduce. Another trait researchers are breeding for is the ability of bees to recognize developing mites inside capped brood cells. These hygienic bees will remove the bee pupae and parasitic mites from the hive. This action does sacrifice the bee pupae but it also removes developing mites. An ideal honeybee colony would have the tendency to do both actions. Researchers continue to strive to breed honeybees better adapted to varroa mite control. Some strains of honeybees are said to be better at mite control than others. However, we await a truly resistant bee. The term has also become a marketing tool to sell bees. You must devise some type of varroa management plan. Yes, even in the first year of the colony. The treatment option chosen will depend on your beekeeping management philosophy.

Integrated Pest Management (IPM) Method

After the influx of varroa mites in the mid-1980s, beekeepers have been at war with varroa. A beekeeping management plan that incorporates natural methods and "soft" chemicals when needed is a common approach. The goal is to keep bees healthy and productive while controlling pests and reducing chemical use. The integrated pest management (IPM) plan consists of mechanical and chemical methods. Soft chemicals are the next step, and approved synthetic chemicals are used last.

Mechanical Mite Control

Mechanical methods of pest control include products such as screened bottom boards. A few years ago, researchers hoped using screened (rather than solid wood) bottom boards would help control varroa mites. The idea was that grooming honeybees would dislodge mites, causing them to fall to the ground and die.

Alas, the percentage of mites that fall from the colony does not seem to be enough to make a difference. Screened bottom boards are no longer thought to be useful for mite control, but they do have some advantages. They allow for good ventilation during hot summers. You may put the cardboard grid in during winter, and use it to evaluate mite infestations during summer. Every screened bottom board comes with a grid board; don't throw it away.

We do not need to rid the hive of every single varroa mite. Our goal is to keep the pest level below the economic threshold. Honeybee colonies can deal with a small pest problem, but colonies collapse when the pest numbers get out of hand.

Chemical Mite Control

When mechanical methods are not enough, and they often are not, we progress to chemical methods. There are many mite treatments available and none are singled out as being the best. Some of them work well some of the time, but none of them are 100 percent reliable all the time. Always read the label and follow the manufacturer's recommendations when using any product in your beehive. Rotating between two or three treatment methods is a sound practice.

Soft chemicals (thymol, formic acid, oxalic acid, etc.) have the characteristics of leaving little or no residue in the hive. Once the chemical has done its job, you will find minuscule or zero traces in the beeswax or honey. Please don't rely solely on YouTube videos for your education on treating bees for varroa mites. Do your research.

More Approved Methods for Pest Management

If soft treatments fail to control mites in your hives, you have another option: products that are approved for use in beehives by the EPA but may be harder on the bees. These approved chemicals have passed tests to prove their effectiveness. Some of these products can leave measurable amounts of residue in wax and/or honey. Is this a big issue for you? That's your decision. Large commercial beekeepers often need to use synthetic chemical methods for mite control. They do not have the resources and time to use the more labor-intensive IPM techniques. Commercial beekeepers cannot lose thousands of bee colonies and stay in business. They need a management strategy that works and does not involve excessive labor costs.

There is no one perfect way to manage bees. Learn the basics and form your own opinions. Whether you are choosing a hive style, type of bee to buy, or a pest treatment, they are your hives and ultimately you will have to make the decisions.

Testing for Varroa Mite Infestations

In the world of beekeeping, we often see two situations involving varroa mites and bees. The colonies that have mites and the colonies that have mites but the beekeeper doesn't think they do. You cannot rely on a visual inspection to determine a mite infestation. Once the beekeeper is seeing mites on bees, things are serious indeed. Yes, you may be able to see a mite on one of your honeybees. However, it is not the mites we see that are causing the most damage, it is the mites inside capped brood cells damaging the next generation of bees. Mite counts provide a reliable "guesstimate" regarding mite levels. Varroa mite levels are higher in later summer and fall than at any other time of year. But this is not the only time that we are concerned about mite levels.

The level of infestation in a colony is called the "mite count." There are several ways to conduct a mite count but keep in mind that both only provide an estimate. Testing provides advice on whether or not you need to treat. But you should always test again a few weeks after treatment to make sure it worked! The two most common varroa sampling methods are: alcohol wash/sugar shake and a mite drop count.

Alcohol Wash/Sugar Shake Varroa Count

The first way to estimate a mite level is with an alcohol wash. Find the brood area in your colony. Because mites reproduce inside brood cells, the nurse bees in the brood nest will have the higher number of mites. Shake about 300 bees from that area into a wide-mouth glass jar. This is about ½ cup of bees. Be careful to not get the queen!) Now, add 70 percent alcohol (or soapy water) and a lid, and shake. Pour the liquid through a small mesh wire to strain out the bees. Then, re-pour liquid through a smaller mesh filter (coffee filter, cotton cloth, etc.). Any mites dislodged from the bees will be visible. Count the mites. Of course, this kills the bees you are using in the sample. It is better to sacrifice 300 bees than to allow thousands to die due to varroa infestation.

For those of us who cannot face killing that jar of bees, there is the sugar shake method. Harvest 300 bees as above, and pour the bees into a wide-mouth glass jar fitted with a mesh lid. Add 1–3 tablespoons of powdered sugar to the jar and gently shake. You may set the jar in the sun for just a minute to increase humidity if needed. Gently shake the sugar out of the jar (through the mesh lid) onto a white surface and count the mites. Release the sugar-coated bees. (Warning: they will be alive but they may not be happy!)

Mite Drop Count

To determine the varroa mite infestation percentage, divide the number of mites by the number of bees. If you find twelve mites: 12 divided by 300 = 4 percent infestation. I am unhappy with anything over 2 percent but anything over 3 percent is cause for real concern.

New beekeepers often shy away from the prospect of measuring out ½ cup of bees. And they may fear that they will accidentally capture the queen causing damage. There is another method that can be used to give some data about the mite levels in your hive. Every day varroa mites and bees share the space inside the hive. In the normal hustle and bustle, some mites will fall. Researchers have determined a guideline for detecting the level of infestation using a sticky board under the hive.

The sticky board method involves putting Vaseline or cooking spray on one side of a white board. This is the white grid board that comes with screened bottom boards but you can use other materials too. The greased sticky board is placed under the screen bottom for twenty-four hours and then removed. The screen prevents bees from getting stuck on the board. If you have solid wood bottom boards, you can purchase a special screen insert to allow testing. After twenty-four hours, inspect the board and count the number of mites you see. (For greater accuracy, you may leave the board in place for three days and divide the number of mites found by three.) How many mites are too many on your drop board?

Most experts recommend a twenty-four hour drop of below thirty mites is best. However, this is a highly contested number because there are so many variables. The number of bees in the hive and the time of year affect drop rate. I recommend a much lower number and treat any hives with a twenty-four hour drop over ten.

Always use approved methods for varroa mite control. This is especially important for new beekeepers. Don't believe everything you hear on YouTube. It is best to leave the experimentation to the researchers.

Small Hive Beetles

The small hive beetle (*Aethina tumida*) originated in Africa and appeared in the US during the 1990s. This shiny black beetle is about the size of a ladybug. Beetles fly into the hive in search of food and a place to lay eggs. As the beetles attempt to lay eggs on the comb, bees chase them around the hive, but our honeybees are unable to sting the hard-shelled beetles.

The adult beetles do no damage inside the hive. However, they will lay eggs on unprotected areas of the comb. The beetle larvae that hatch destroy comb and ruin honey. Hive conditions can become so bad that the bees will abscond or leave the hive. Small hive beetles can quickly destroy weak colonies. Even strong colonies are not completely safe from beetle destruction.

As a mechanical method of control, use beetle traps to catch hive beetles. This is not a perfect solution but it is a non-chemical strategy to reduce beetle numbers. One popular beetle trap is the "beetle jail." This plastic trap hangs between two frames inside the hive. Mineral oil is placed in the open compartments of the trap. When beetles scurry inside to hide from bees, they become trapped. There are several types of beetle traps and trays on the market. Some of them work better than others and none are a clear favorite.

Another mechanical method for small hive beetle control is to squash any beetle in sight with your hive tool. This may seem a little harsh until you have seen how small hive beetles can destroy a hive. If the small hive beetle

Small hive beetles

has not made it to your part of the world, you are one lucky beekeeper. In the south, beekeepers battle them constantly. Few chemicals are approved for beetle treatment, and the one that is approved (Checkmite+) has detrimental effects on drone bees and is not used by most beekeepers. The best line of defense against small hive beetles is to keep strong colonies in sunny locations and use traps as needed.

Wax Moths

A wax moth is a flying insect that is most active at night. There are actually two types of wax moths that trouble beekeepers. The greater wax moth (*Galleria mellonella*), and the lesser wax moth (*Achroia grisella*). Both types of moths are attracted to the hive by odors and often enter the beehive a night. The adult moths are the beginning of a possible disaster for a hive that does not have a sizable population. Yet, as with hive beetles, adult moths are not the real problem.

If the population of bees in a hive is not large enough to guard all of the comb, adult moths lay eggs. Moth eggs can hatch in three days during warm conditions. These wax moth larvae can cause a large amount of damage. When wax moth larvae hatch, they are very small white grubs. Then, they begin to eat and grow. Many beekeepers put out "moth traps" to try to lessen the number of adult moths entering beehives. These traps are commonly homemade with many different "bait" recipes in use. Perhaps you can experiment with bait recipes. But don't expect traps to be the complete answer to moth problems.

Wax moth larvae eat beeswax, the remains of bee larval cocoons, bee cocoon silk, and bee feces in the cell. All of these are present in older frames of comb. Bee cocoons are the materials left behind by a developing honeybee. Any comb that has had bee brood will be more attractive to wax moths. This preference for used comb is why wax moth larvae problems develop near the brood nest of the hive. But wax moth larvae can live on pure beeswax. Therefore, comb from stored honey supers is not completely safe.

Older moth larvae turn gray and can measure up to twenty-eight millimeters in length. They are very mobile and can move from one hive to another, although this doesn't happen often. They spin white cocoons for the transformation into adults. They will often eat away at wooden surfaces in the hive, creating a wavy surface and causing damage to parts of the beehive.

Do wax moths kill beehives? No, not really. A wax moth infestation is an indicator of another problem in the hive. A strong colony of honeybees will keep wax moth infestations under control. Do you have a lot of bees in the hive? Are there enough bees to cover most of the comb surface? The hive must have a large enough population to prevent adult moths from laying eggs. It is the weak beehives that are at greatest risk for moth damage. If your hive is weak, you must try to learn why. Perhaps, something has happened to your queen causing hive population to plummet. A queen failure or any condition that allows the bee population to drop needs attention. If the beekeeper puts too many boxes on a small, struggling colony of bees, wax moths can enter the hive and infest the empty comb.

Wax moth larva and hive damage

How does a beekeeper know if they have a problem with wax moths in the hive? Seeing an adult wax moth fly out is not a cause for concern. But you may see moth larvae during hive inspections. Wax moth larvae can be found crawling on the comb surface. (They are easily confused with small hive beetle larvae, another bee pest.) The defining characteristic of wax moths is the presence of webbing. Wax moth larvae tunnel through the honeycomb. They leave behind noticeable tunnels with a "spider-like" webbing. This process continues until all the wax has been consumed, leaving a webby mess for the beekeeper to clean. Their feces (small cylindrical black pieces) can also be seen on the bottom board.

The best defense against wax moths is to maintain strong colonies. A healthy, strong colony will repel an attack and keep the beehive relatively moth-free. Don't give your honeybees more space than they can patrol. Too much space + too few bees = big trouble. Take extra boxes off the colony when not needed. Leaving too many honey supers on the bee colony in late summer is tricky. If the bee population declines, you could end up with small hive beetles and wax moth larvae in the hive. Monitor the population of your colonies and check colonies that you know have swarmed. If the population has dropped dramatically, you have comb that cannot be patrolled. Now, adult moths can move in to lay eggs.

Chapter 11

Enjoying the Bounty of the Hive

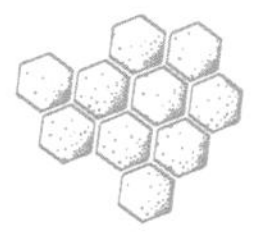

Beekeepers understand more than anyone the effort required to produce honey and beeswax. Because honeybees are so industrious, they are capable of making enough for themselves and us as well. After all this hard work, it is only natural that we would want to get every bit of value out of these products. One of the major perks of beekeeping is having all this wonderful honey and beeswax to use.

Ways to Use Honey

Honey is a natural sweetener and can be used in many ways. In addition to eating honey raw because it tastes good, it has also been used as a cooking ingredient for a long time. Cooking with honey does cause the loss of some of its wonderful raw properties. In spite of that, it adds flavor, sweetness, and moisture to baked goods. One of the best choices for long-term food storage, honey is safe for years and years when protected from moisture in a sealed container.

With just a few hives, the crop of extracted honey can be stored in sealed quart jars and used as needed. Larger honey producers use food-grade five-gallon buckets with tightly sealing lids. When choosing a storage method for your honey crop, remember that honey is heavy. A five-gallon pail of honey weighs at least sixty pounds.

Another factor to consider is honey crystallization. Almost all honey will crystallize over time. This natural process of honey does not mean it is bad. Crystallized honey can be gently warmed and it will return to its liquid state. Glass jars are easy to warm in a hot water bath. Crystallized honey can be scooped from a plastic bucket.

What else can we do with honey? Thousands of people swear that raw honey helps ease the discomfort of seasonal allergies. Though no known research verifies these claims, advocates continue their raw honey regime. The most common method of use is to take 1 tablespoon of raw honey per day year-round. This can be added to coffee or tea if the liquid is not very hot. Medical studies do not support honey as a treatment for allergy sufferers, but don't tell that to the folks who swear that it works!

Honey for Coughs and Sore Throats

Honey is known as an effective treatment for coughs. Even modern medical practitioners recommend the use of honey to ease coughing. Do remember that raw honey or any raw food should not be given to infants without doctor supervision. The next time you have a cough or sore throat, reach for your jar of honey. Make a cup of very warm tea or water, add 1 tablespoon of lemon juice and 2 tablespoons of honey, and stir. Sip slowly while warm.

Honey for Wound Care

The medicinal benefits of honey do not stop at allergy relief or cough and cold treatments. We know that raw honey has antiseptic and antibacterial properties. Honey has been used in ancient battlefields as a wound treatment for ages. Honey can be applied under a bandage for cuts and other wounds. If you have wounds on or near your lips, a small smear of honey is preferable to toxic medicine. A small container of honey is an excellent take-along while hiking. But don't put on too much honey if you are in bear country!

Honey for Beauty

The antiseptic and moisturizing properties of raw honey have a place in many beauty regimes. A popular face washing technique involves spreading a thin layer over dry skin, letting it sit for two minutes, and rinsing well. People who use this technique claim that the treatment refreshes and moisturizes their skin.

Do you have a problem with dry scalp? You might try a common honey and coconut oil scalp treatment. Participants make a paste of honey and coconut oil, apply it to scalp, and let sit for five minutes, then wash hair as normal.

The skin-protecting properties of honey and beeswax is why they are popular additions to soaps and lip balms. Natural products are often just as effective as those with man-made additives. In addition, they contain fewer questionable chemicals. Seriously, do we want to put things on our bodies that have unpronounceable names?

Beeswax: A Valuable Treasure from the Hive

Beeswax is produced by honeybees and used to create sheets of honeycomb. When beekeepers harvest honey, surplus beeswax is left over from the extraction process. This extra beeswax is not thrown away. It is melted into blocks and stored to be used later.

Have you ever held a piece of natural beeswax? It feels almost magical, and it smells great too! Beeswax is used in its raw form for many projects, and it can be mixed with other ingredients to create products that add value to our lives. The many uses for raw beeswax go way beyond making beeswax candles. Beeswax plays a role in our health care, beauty routines, home life, in the workshop, and more.

Uses for Beeswax

Do you have a problem with sticky drawers? No, seriously. I am of course referring to the pull-out drawers of your wooden furniture. Beeswax is the answer. Take a small block (piece of beeswax) and rub it on the runners of the bottom of the drawer. This should provide enough lubrication to permit the drawer to slide easily.

Many seamstresses know the secret of using beeswax for easier sewing. Everyone needs to sew on a button or two at some point in their lives. You

Fresh beeswax

only need a small block of pure beeswax to make the task easier. Cut your thread to the length needed. Wrap the thread around the block of wax and firmly slide it along the surface. Push down with your finger if needed and you should have a light coat of wax on your thread. Waxed thread is easier to work with and less likely to knot.

Do you have a favorite coat with a zipper problem? Sticky zippers can be so annoying. One of my beekeeper suits is prone to having a difficult zipper. When this happens, I use a small piece of beeswax to rub along the teeth of the zipper. Works like a charm, and it will work for your favorite coat, too.

Due to its adhesive properties, beeswax can be used to seal envelopes. This sounds like such a neat idea to do, especially during the holidays. The mail would smell very good!

Creating a furniture polish has been one of the most popular uses for beeswax for hundreds of years. The traditional recipe involves mixing equal parts of turpentine, boiled linseed oil, and melted beeswax. Make sure your beeswax is free of dirt, etc. You do not need to heat this mixture. Once the ingredients are combined in a glass jar, mix well and let it sit for a day or two. After it thickens you have a great natural furniture polish. Be sure to buff it out within ten minutes for a nice shine. And always test any product on a small out-of-sight spot first!

Raw beeswax makes a good polish for bronze pieces, too. Rub a warm beeswax bar (you can warm it with a hair dryer), on your bronze items, and buff. The wax coating will prevent tarnishing.

You can even use beeswax in the kitchen. Beeswax has been used for hundreds of years in canning and preserving food. Though not as popular today, it was once used to prevent spoilage of jam or jelly. Once the jam was finished, a thick coat of wax was applied directly over the food. I still do this if I have half a jar of jam and don't want to bother with the lid. (I must admit that I also put it in the refrigerator because I am not as fearless as Grandma was in her day.)

Our culture is finally getting the message about our exorbitant use of plastics. All this plastic has to go somewhere once we are finished. Instead of adding to our ever-growing landfills, consider making your own reusable beeswax wraps. Small amounts of melted beeswax, olive oil, and pine resin are melted and brushed on precut pieces of cotton cloth. After cooling, they can be used again and again as dish covers. They are all natural and can be used without worry as they contain no toxic chemicals.

A small amount of beeswax can be rubbed inside your favorite frying pan or sauce pan. The pan should be slightly warm to aid in wax transfer. Buff gently and you will have a natural nonstick surface for cooking.

You don't have to limit your use of beeswax to candles and inside projects. A small block of beeswax will come in handy in your garage or shop. In fact, I have sold many small blocks of beeswax to folks who give them as small gifts!

When using a handsaw in new wood, any carpenter will appreciate having a little beeswax on hand. By rubbing the wax on the teeth of the saw, it will cut through the wood much easier. (This is the same principle as the sticky drawer problem we talked about earlier.)

Keep a small block of wax in your shop near the wood screws. Trying to insert a wood screw into a hard piece of old wood can be difficult even with an electric drill driver. Rubbing the tip of the screw over the wax block makes the job a lot easier.

One of the often overlooked uses for beeswax is as a water-proofer. Do you have an older pair of work boots? Maybe they still have a little use in them but they leak in rainy weather? I hate wet feet! Melted beeswax may be the answer and help you get a few more steps out of those boots. Melt beeswax and brush a light coat over the surface of the leather boots. You can apply extra to the seams and other troublesome areas.

If you enjoy backpacking or camping, you can use beeswax to make lightweight fire starters. Brush melted beeswax onto small squares of cardboard. Let cool and pack. These are easy to carry, don't take up much space, and will help you start your next fire.

Let's make beeswax lip balm. Okay, it's not only beeswax. You can imagine that dragging a block of wax across your lips won't work. Still, the beeswax plays an important part in our recipe. It protects and moisturizes your skin. When I use beeswax to create lip balm, for one recipe I mix ⅓ beeswax, ⅓ sustainable palm oil, and ⅓ sweet almond oil. Melt all ingredients and combine. Then you can pour the warm liquid into lip balm tubes or pots! You can add a few drops of peppermint essential oil if you wish to give your beeswax lip balm a buzz.

Cracked heels can benefit from a beeswax inspired cream. I mix ½ cup of coconut oil, ½ cup of shea butter, and 1 ounce of pure beeswax in a container. Stir well. You can add some magnesium flakes if you wish (optional). Once the cream has set, it should be stored in the refrigerator. Rub the beeswax cream on your heels at night and put on a pair of socks.

A mixture of ½ beeswax and ½ coconut oil is often used as a hair dressing. It makes a great beard balm for the guys and can even be used to condition dreadlocks.

Conclusion

The world of beekeeping is one of continued amazement and wonder. A beekeeper will never learn everything there is to know about bees. Bees have some secrets that they keep to themselves. With hard work and patience, your beekeeper skills will grow. Enjoy your bees and don't forget to have fun. Do not be too hard on yourself when you lose hives. It will happen. As your knowledge grows, you will come to realize how little we really know about bees. The bees are in charge—not the beekeepers. Come, join me on the journey known as beekeeping. It can be a wild ride, but I love it. And I think you will too.

Resources

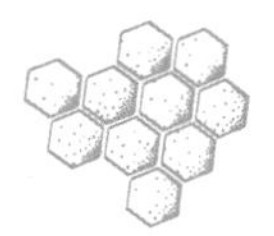

Where to Buy Bees and Supplies

Most major bee suppliers sell bees during the spring and beekeeping supplies year-round. The bee ordering process begins in late winter for spring delivery. Purchase from a company with a good reputation for delivering healthy packages. You may also find beekeeping supplies for sale at various other retailers like the ones listed below.

Kelly Beekeeping
www.kelleybees.com

Mann Lake, Ltd
www.mannlakeltd.com

Rossman Apiaries, LLC
www.gabees.com

Beekeeping Education Resources

The Honeybee Conservancy
thehoneybeeconservancy.org/education

USDA-ARS Carl Hayden Honeybee Research Center
www.ars.usda.gov

Scientific Beekeeping
scientificbeekeeping.com/the-rules-for-successful-beekeeping

The Old Farmer's Almanac
www.almanac.com/news/beekeeping/beekeeping-101-why-raise-honey-bees

Penn State Extension
extension.psu.edu/beekeeping-honey-bees

Acknowledgments

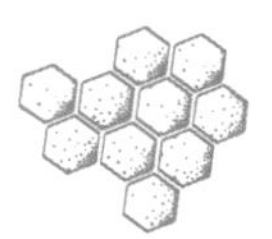

I am eternally grateful to my husband who encouraged me to reach for my dreams and supported my beekeeping adventures in so many ways.

Heartfelt appreciation goes out to all those beekeepers who shared and continue to share their love of bees with new beekeepers.

A special thank you to the Skyhorse Publishing team who made this book possible. And especially to Nicole Mele who said "you can do it." She was right.

Index

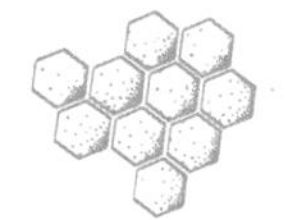

A

abdomen, 13
adult bee, 16
Africanized honeybees (AHB, killer bees, *Apis mellifera scutellata*), 14
air flow in hive, 28
alcohol wash, 80–81
allergy to bees, 11–12
  developing, 11
American foulbrood, 32
apiarist, xiv
apiary, 47–51
Apiguard, 6
Apis mellifera, 14
Apistan (fluvalinate), 5
Apivar (amitraz), 5
arrival of bees, 53–62. *See also* Installing bees
  feeding, 54, 58–59
  shaking them in, 55–57
  slow release, 57–58
  temperature, 53–54

B

bee. *See also* Arrival of bees; Bees, getting; Death; Feeding bees; Installing bees; Mating; Reproduction
  abdomen, 13
  arrival, 53–62
  body sections, 13
  bread, 20
  buckfast, 14
  Carniolan bees *(Apis mellifera carnica),* 14, 18
  Caucasian bees *(Apis mellifera caucasica),* 14
  colony. *See* Colony
  colors, 15
  family tree, 14
  feeder, 41, 58
  female, 18, 23
  fever, 31
  flight activity, 64
  glue, 45
  honey stomach or crop, 18, 20, 27, 74
  house, 20
  hybrid, 14
  hygienic, 14
  killer. *See* Africanized honeybees (AHB, killer bees)
  legs, 13
  life cycle, 16
  lifespan, 1, 13, 24
  male, 16
  mandibles, 27–28
  math, 37
  navigation, 47
  nurse, 19, **19**
  proboscis, 27
  purebred, 14
  races, 14
  robbing, 64
  Russian, 18
  scout, 74

species, 14
syrup, 41
tails, 17
thorax, 13
wings, 13, 69
Bee Culture, 3
beehive. *See* Hive
beekeepers, experienced, 3
beekeeping
beginning, xiv
business, xiv, xv
cost, 32
experience, 2
philosophy, 4
physical demands of, 8
popularity, xi
reasons for, 3
time, 7–8
treatment-free, 6
beekeeping supplies, 38–43
used *vs.* new, 32
bees, getting, 34–38
established hive, buying, 34, 37
nuc (nucleus) of bees, buying, 34, 36–37
package of bees, buying, 34–36, **35**
swarm, catching, 34, **37,** 37–38
beeswax, 19, 28, 66, 84, **90**
beards, 92
boots, 92
bronze, 91
candles, 91
canning and preserving food, 91
comb, 32
drawers, sticky, 89
envelopes, 90
fire starters, 92
frying pan, 91
furniture polish, 91
hair, 92
heels, cracked, 92
lip balm, 92
saw, 91–92
sewing, 89–90
uses, 89–92
waterproofing, 92
wood screws, 92
wraps, 91
zipper, sticky, 90
beetle, 48. *See also* Small hive beetles *(Aethina tumida)*
control, 82–83
jail, 82
squashing, 82
trap, 82
birth, 21. *See also* Mating; Reproduction
boxes, 33
brood, 36, 60
capped, 66
food glands, 18–19
milk, **65,** 66
pattern, **61,** 65–66, 69
stages, 65
buckfast bees, 14
bumblebees, 28
business of beekeeping, xiv, xv

C

cage. *See* Queen, cage
candy, 35, **54,** 55, 57, 60
Carniolan bees *(Apis mellifera carnica),* 14, 18
Caucasian bees *(Apis mellifera caucasica),* 14
cell capping, 66
challenges for new beekeepers, 7–11
Checkmite+ (coumphos), 5, 83
chemical treatments, 4–5, 79–80
soft, 5–6, 78–79

classes about beekeeping, xiv
climate, 3
clubs, xiv
colony, 2, 13–25
cooperative brood care, 13
lifespan, 1
overlapping generations, 13
queenless, 67
reproductive division of labor, 13
stress, 3
combs, 36, **40**
building comb (drawing comb), 68
burr comb, 68
cross comb, 68
spacing, 33
used, 32
communication, 24, 45

D
dampness, 51
daylight, 15
death, xii, 18, 23, 55, 60, 70
decline of pollinators, xi
defensiveness, 64
dehydration, 27
diseases, 32, 37, 77–86
drone bees, 15–18, **17,** 35, 67
brood cells, 66–67
death, 18, 23
eyes, 17
mating, 23
mating flights, 18
maturity, 18
number of, 17
reproduction, 18
size, 17
stingers, 17
tails, 17
drone congregation area, 23
drought, 7

E
eggs, 16, 23, 60, **61,** 65
Environmental Protection Agency (EPA), 4, 79
enzymes, 27
eyes, 17, **17**

F
feeding bees, 7, 41, 63
on arrival, 54, 58–59
sugar water mixture, 58
winter, 69
flight activity, 64
foraging, 21, 41
formic acid, 6
foundation, 39
frames, **24,** 32–33, 36, 38–39, 41, 45, 56–59, 63
lifting, 8
moving, 59–60
used, 32

G
gluconic acid, 27
glucose oxidase, 27
glue, 41
goals, defining, 1–6

H
head, 13
hive, **xiii**. *See also* Arrival of bees
access, 51
air flow, 28
bees in, 15
configuration, 39–40
entrance, 41
fully established, 37
and light, 64
location, 48, 51, 70
number of, 3, 15, 31
population, 15, 21, 86
problems in hive, recognizing, 8
stand, 48

temperature, 51
tool, 43, 45, 55
ventilation, 79
The Hive and the Honey-Bee (Langstroth), 33
hive components, 38–42
assembling, 41
bottom board, 39
bottom board, screened, 78–79
combs, 39
entrance reducer, 40
foundation, 39
frames, 39
hive body, 39
inner cover, 39
super, 39
telescoping top, 39
hive inspection, 63
brood pattern, 65–66, 69
brood stages, 65
comb, 68
and disruption, 64
drone brood cells, 66–67
frames, 68
frequency, 64
goals, 64
honey, 69
laying workers, 67–68
mites or pests, 69
open hive, 69
pollen, 69
protective gear, 64
queen, 63, 65, 69
schedule, 64
smoke, 69
techniques, 64–65
time, 64
tips, 69
what to look for, 65–69
hive management, xiv
frequency of, 60
new hive, 59
hive style, 32–34
choosing, 34
Langstroth hive, 32–34, 38
top-bar hive (Kenyan hive), 33–34
honey, 1, 69
and allergies, 88
amount of, xii
antibacterial properties, 27, 88
beauty regime, 88–89
bees eating, **2**
capping, 28, **28**
cooking with, 87
coughs and sore throat treatment, 88
crystallization, 88
drying process, 28
extractor, 2, 33
flow, 29
harvesting, 29, 41, 70, **71**
harvesting, crush-and-strain, 33, **33**
left on hive for winter, 8
lips, 88–89
moisture content, 28
pots, 28
production, 27–29
raw, 87
scalp, 89
skin, 88–89
spoilage, 28
stability of, 28
stomach or crop, 18, 20, 27, 74
storage, 28, 87
sweetener, 87
ways to use, 87–89
weight, 87
when bees make honey, 28–29
wound care, 88
honeybee, type of, 3

honeycomb, 89
Hopguard, 6
house bee, 20
hybrid bees, 14
hydrogen peroxide, 27
hygienic bees, 14, 78
hypopharyngeal glands, 27

I

infestation, 80
installing bees, 41
integrated pest management plan (IPM), 78
invertase, 27
investment, xiv
Italian bees *(Apis mellifera lingustica),* 14, 18

K

killer bees. *See* Africanized honeybees (AHB, killer bees)

L

Langstroth, Rev. L., 32–33
larvae, 13, 16, 60, 65–66, 74
  queen, 25
learning from experienced beekeepers, 3
legs, 13, 19
life cycle, 16
lifespan, 1, 13, 24, 59
limits, knowing your, 31

M

male bees, 16
management style, 3–4
mandibles, 27–28
mating, 16, 23, 25. *See also* Reproduction
  flights, 18
medications, 32
mite control
  chemical, 79–80
  mechanical, 78–79
mite counts, 80
  alcohol wash, 80–81
  mite drop count, 81–82
  sticky board method, 81–82
  sugar shake, 81
mites, 69. *See also* Varroa mites *(V. destructor)*
  and hygienic bees, 78
moths. *See* Wax moths

N

navigation, 47
nectar, 15, 21, 27
  sharing, 27
  transfer of, 20
number of bees in hive, 3, 15, 31
nurse bee, 19, **19**

O

odors, 51
overheating, 35
overwhelmed feeling, 8
overwintering. *See also* Winterizing
  boxes for, 63
  inspection, 64
oxalic acid, 6

P

package of bees
  arrival of, 53–59
  buying, 34–36, **35**
  cork plug, 56–57
  dead queen, 55
  dry package, 55
  feeding, 58–59
  installation, 55–58
  queen, 54, **54,** 55
  queen cage, 54, **54,** 55–57

shaking them in, 55–57
slow release, 57–58
syrup can, 56
temperature, 35, 53–54
parthenogenesis, 16
pest control, 78–80
cardboard grid, 79
chemical mite control, 79–80
integrated pest management plan (IPM), 78
mechanical mite control, 78–79
pesticide, xii
pests, 1, 3–6, 69
pheromones, 23–24, 57, 74
alarm, 9–10
philosophy, 3–6
physical demands of beekeeping, 8
poison gland, 21
pollen, 20, **20,** 21
baskets, 18–20
depositing, 20
stored, 69
pollination, 1
popularity of beekeeping, xi
population of hive, 15, 21, 86
preparation, 51
problems in hive, recognizing, 8
proboscis, 27
propolis, 21, 45
protective gear, 10, 32, 42, 64
beekeeping jacket, 42
clothing, 42
gloves, 42
hat and veil, 42
suits, 42
pupa, 16, 66, 78
pure-bred bee, 14

Q

queen, 13, 15–16, **22,** 22–25, **24,** 35–36, 63, 65, 74
adding, 67
aging, 25
arrival, 54, **54,** 55
cage, 54, **54,** 55–56, 59–60, 67
cell development, 64
cells, 25, 67, 74–75
dead, 55, 60
development, 22
failed, sign of, 67, 84
hatching, 25
how a queen is made, 24–25
larvae, 25
lifespan, 24
marking, 62
mating, 25
number of, 23
pheromones, 23–24, 74
queenless colony, 67
release from package, 59–60
replacing older, 75
reproduction, 23
re-queening, 25
right, 66
rivals, 23, 25, 74
room to lay, 75
seeing, 69
size, 23
spermatheca, 23
status, 59–60, 62
stinger, 23
and swarms, 37
verifying, 63
virgin, 23, 25

R

races of honey bees, 14
reproduction, 13, 18, 23. *See also* Mating

cycle, 59
and swarming, 73
re-queening, 25
robbing bees, 64
rot, 49
royal jelly, 25
Russian bees, 18

S

saliva, 20, 27
scout bees, 74
semen, 23
skunks, 49
small hive beetles (*Aethina tumida*), 82–83, **83**. *See also* Beetle
beetle jail, 82
squashing, 82
traps, 82
smoker, 10, 43, **44,** 45, 69
accelerants, 45
fuel, 43
lighting, 43
using, 45
social insects, 13
species of bees, 14
sperm, 67
spermatheca, 23
spring, 64
starvation, 59
stinger, 17, 21, 23
barbed, 18
stings, 9, 42
getting stinger out, 9
reaction to, 9–11
reason for, 9
reducing, 10
sugar, 27
feed, 32, 41
shake, 81
summer, 63
swarm, 15, 34, 37, **37,** 37–38, 58, 64, 73–75
defined, 73
honey, 73
and honey production, 38
honey stomachs, 74
management tips, 75
mode, 74
new colonies, 73
new queen, 74
planning, 73–74
prevention, 73
queen, 25, 73
and queen age, 25, 74
reducing, 38
and reproduction, 73
season, 74
signs of, 64–65
and space, 74
temporary resting place, 74
throw a swarm, 74
time, 74
traps, 75
in a tree, 74
triggers, 73
sweating, 8–9

T

tails, 17
tasks, progression of, 21
taxonomy, 14
temperature, 35–36, 50–51, 53, 70, 74
thorax, 13
time for beekeeping, 7–8
time requirement, 7–8
tools, 43
travel with bees, 35–36, 53
treatment-free beekeeping, 6

V

varroa mites (*V. destructor*), 4, **5,** 63, 77, **77,** 78. *See also* Mites
developing mite-resistant bees, 77–78
feeding activity, 77

and hygienic bees, 78
integrated pest management plan (IPM), 78
management plan, 69
philosophies about, 4–6
testing for infestation, 80–82
venom, 10
ventilation, 79

W

water, 21
access to, 49–50
source, 49–50, **50**
wax. *See* Beeswax
wax glands, 18–19
wax moths, 83–86, **85**
cocoons, 84
defense against, 85–86
feces, 85
greater wax moth *(Galleria mellonella),* 83
and hive weakness, 84
larva, 85, **85**
lesser wax moth *(Achroia grisella),* 83
prevention, 85–86
recognizing a problem, 85
traps, 84
webbing, 85
weather, 8, 29
and bee temperament, 10
patterns, xii, 8
protecting hive from, 51
webbing, 85
wind, 51
wings, 13
unhooking, 70
winter cluster, 69
winterizing, 69–70
feeding, 69
heat management techniques, 70
honey, 69
temperature, 69–70
when to begin, 69
winter survival, 15, 21
worker bees, 13, 15–16, 18–22, **19,** 35, **66**
barbed stinger, 18, 21
body parts, 18
brood cells, 66–67
brood food glands, 18–19
death, 21
fertilized eggs, 23
foraging, 27
honey stomach or crop, 18, 20, 27, 74
laying workers, 67–68
legs, 19
lifespan, 21–22
numbers of, 18
poison gland, 21
pollen baskets, 18–20
proboscis, 27
role of, 21–22
seasonal changes, 22
wax glands, 18–19

Notes

Notes

Notes

Notes

Notes

Notes